War, Peace, and
All That Jazz

A   H I S T O R Y   O F   U S

Oxford University Press

OXFORD
A HISTORY OF
US
BOOK NINE

War, Peace, and All That Jazz

Joy Hakim

Oxford University Press
New York

Oxford University Press

Oxford   New York

Athens   Auckland   Bangkok   Bogotá
Bombay   Buenos Aires   Calcutta   Cape Town
Dar es Salaam   Delhi   Florence   Hong Kong   Istanbul
Karachi   Kuala Lumpur   Madras   Madrid
Melbourne   Mexico City   Nairobi   Paris   Singapore
Taipei   Tokyo   Toronto   Warsaw

and associated companies in
Berlin   Ibadan

Designer: Mervyn E. Clay
Maps copyright © 1995 by Wendy Frost and Elspeth Leacock
Produced by American Historical Publications

Published by Oxford University Press, Inc.

198 Madison Avenue, New York, New York 10016

Oxford is a registered trademark of Oxford University Press

Library of Congress Cataloging-in-Publication Data
Hakim, Joy.
War, peace, and all that jazz / Joy Hakim.
p.   cm.—(A history of US: bk. 9)
Includes bibliographical references and index.
ISBN 0-19-507761-X (lib. ed.)—ISBN 0-19-507765-2 (series, lib. ed.)
ISBN 0-19-507762-8 (trade paperback ed.)—ISBN 0-19-507766-0 (series, trade paperback ed.)
ISBN 0-19-509514-6 (trade hardcover ed.)—ISBN 0-19-509484-0 (series, trade hardcover ed.)
ISBN 0-19-511079-X (school paperback ed.)—ISBN 0-19-511070-6 (series, school paperback ed.)
1. United States—History—1901–1953—Juvenile literature.
[1. United States—History—20th century.] I. Title. II. Series: Hakim, Joy. History of US; 9.
E178.3.H22 1994 vol. 9
[E741]
973.91—dc20   93-28768
CIP

7 9 8
Printed in the United States of America
on acid-free paper

"What happens to a dream deferred" (page 6) is by Langston Hughes. Copyright © 1951 by Alfred A. Knopf, Inc.. Reprinted by permission of Alfred A. Knopf, Inc. The poem by Ogden Nash on page 6 is reprinted by permission of Little, Brown, Inc. The passage on page 15 is from *A Girl from Yamhill* by Beverly Cleary. Copyright © 1988 by Beverly Cleary. Reprinted by permission of Morrow Junior Books, a division of William Morrow, Inc. The excerpt on page 56 from *Having Our Say: The Delany Sisters' First Hundred Years* by Sarah and A. Elizabeth Delany with Amy Hill Hearth is reprinted with permission of Kodansha America, Inc. Copyright © 1993 by Amy Hill Hearth, Sarah Louise Delany and Annie Elizabeth Delany. The passage on page 67 is reprinted with permission of Scribner's, an imprint of Simon & Schuster, from *The Spirit of St. Louis* by Charles A. Lindbergh. Copyright © 1953 by Charles Scribner's Sons; copyright renewed © 1981 by Anne Morrow Lindbergh. The passage on page 114 is from *Night* by Elie Wiesel. Copyright © 1960 by MacGibbon & Kee, renewed © 1988 by the Collins Publishing Group. Reprinted by permission of Hill & Wang, a division of Farrar, Straus & Giroux, Inc. The passage on page 145 is from *Farewell to Manzanar* by James D. and Jeanne Wakatsuki Houston. Copyright © 1973 by James D. Houston. Reprinted by permission of Houghton Mifflin Co. All rights reserved. Ernie Pyle's column "The Death of Captain Waskow" on page 156 is reprinted by permission of the Scripps Howard Foundation.

IDA GINSBURG FRISCH, who played on the first girls' basketball team in Glens Falls, New York, was a flapper who bobbed her long hair and wore a coat with a mink collar.

She was hardly beyond her teens when she won an automobile for selling more subscriptions to the GLENS FALLS POST-DISPATCH than anyone else in town. (Having a car was unusual when she was a girl.) She didn't know how to drive, but a license and instructions came with the car, so she convinced her sister, Libbie, to get in with her, and the two of them took off for Saratoga. Fifty years later, Libbie still remembered that drive. "I didn't think I'd live through it," she said.

When Ida died, in her eighties, she was the youngest person I knew. She never stopped surprising me with her wit and vitality and curiosity. "What will happen if we try this?" "Or that?" "And let's see for ourselves." I couldn't keep up with her. She was my mother.

JOHN MICHAEL FRISCH—"Jack" to those who knew him—was shy and courtly and had strawberry-blond hair. He came to this country at age three, from Zhitomir in Ukraine, the son of poor Jewish immigrants. Soon there were two sisters. Before Jack finished school, his father had died and Jack had to support the family.

Photographs from the Roaring Twenties show him as a dandy who went to Broadway openings and wore spats on his shoes. I knew him later, when he sat in a big chair, read newspapers and books, thought about things, and knew the answer to any question my schoolbooks asked.

He always said exactly what he meant, and everyone understood that he was a man who could never be unkind or untruthful.

He taught me to love ideas and words and to listen to the music in language. Besides that, he could do magic tricks and stand on his head. He was my father.

We Americans today—all of us—we are characters in the living book of democracy. But we are also its author. It falls upon us now to say whether the chapters that are to come will tell a story of retreat or a story of continued advance.

—FRANKLIN DELANO ROOSEVELT, 1940

What happens to a dream deferred?

Does it dry up
like a raisin in the sun?
Or fester like a sore—
And then run?
Does it stink like rotten meat?
Or crust and sugar over—
like a syrupy sweet?

Maybe it just sags
like a heavy load.

Or does it explode?

—LANGSTON HUGHES

One thing about the past,
It is likely to last.
Some of it is horrid and some sublime,
And there is more of it all the time.

—OGDEN NASH

Clockwise from top left: FDR reports to Congress following the Yalta Conference, March 1945; Babe Ruth in eBULLient mood; "A shack for negroes only at Belle Glade, Florida," April 1945.

Contents

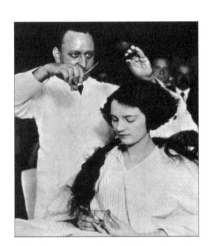

Growing up in the 1920s: Bernice bobs her hair. (Look up F. Scott Fitzgerald at the library for a great short story with that title.)

My hair!! how the wind blows.

ROCKET LINE.

PREFACE
Time Travelers

Benjamin Franklin looking serious and scholarly when he was ambassador to France, around 1780.

Thomas Jefferson

James Madison

George Washington

How about climbing into a time capsule with me? You can set the ship's dial back a bit more than two centuries—to 1789. We're on our way to pick up some old friends: Benjamin Franklin, Thomas Jefferson, James Madison, and George Washington. Let's show them what is happening to the government they worked so hard to found. They called that government an "experiment"; they'll be curious to know how the experiment turned out.

Do you see them there, looking just as they did when we were with them last? You might want to help Ben climb on board; his gout is bothering him. Now, set the dial at 1917, which is a good year to begin this trip in time. Do you wonder what the Founding Fathers will think of these United States? Will they be startled by what they see?

Well, we've put old Ben in Philadelphia, and there he is—right where you would expect him to be—trying out a telephone. We'll never get him away from the new gadgets. Look at him now! Ben's dancing a jig, he's so excited about electric lights. You can't blame him; after all, it was his experiments with electricity that helped that branch of science get started.

But what about George Washington, Thomas Jefferson, and James Madison? We set them down in Washington, D.C. What do they think of the American government? Are they pleased? Are they surprised?

Yes, they are pleased—and not too surprised. The government is running much as they planned it. Power is divided between the states and the federal government. That federal government (in Washington,

9

Jefferson entered, and lost, the competition to design the Capitol (above, looking toward the Washington Monument).

By the 1930s the radio was an important piece of furniture.

D.C.) hardly seems to touch most people's lives. It is only through the post office that it reaches the average citizen. The Founders learn that a just-approved amendment to the Constitution will allow Congress to impose an income tax. James Madison says that will certainly make Americans more aware of the government in Washington.

Jefferson is pleased that most Americans still live on farms or in very small towns—it is the way he meant America to be. And he is delighted with the elegance and beauty of the capital city. (You remember that he helped with its planning.) He says that Pierre L'Enfant's plan for the city seems to have turned out splendidly.

George Washington likes the president, Woodrow Wilson. After all, Wilson is a Virginian, and dignified, and a scholarly man. They all sit down together and discuss politics and philosophy.

Now that our friends have started this journey through time, they can't wait to go on. So let's capsule on—to 1990.

Watch Ben Franklin! He may get run over. His head is in the sky looking at high buildings and airplanes. Wait until he hears a radio. And sees TV!

"Astounding how healthy everyone seems," says Ben when he finally looks down. "Doesn't anyone suffer from the gout?"

"Have you ever seen smiles like these?" asks George Washington, pointing at some boys and girls with nice white teeth.

But look at Thomas Jefferson. He's acting strangely. Someone is telling him about all the century's wars. Someone is telling him about the huge federal and state governments. About the giant American military forces—army, navy, air force, and marines. Someone is telling him that most Americans live in big cities, not on farms. Jefferson looks ill.

Someone is telling George Washington about presidents who have thousands of people working on their staffs. Washington says it sounds kinglike. He calls it "an imperial presidency." So do other people.

The Founding Fathers say this isn't what they intended for the United States. They didn't want standing armies. They didn't want big government. They hate bureaucracies. Ben groans. "It seems like the cumbersome government of George III," he says.

Someone explains that it was all those 20th-century wars—and a depression, too—that made big government necessary. Actually, it began with that man they liked so much, President Woodrow Wilson. He needed to fight a war; things had to be done quickly; so he asked Congress for special powers, war powers.

Well, once one president had started things, it got easier to ask again, and again. First the presidents, then Congress, then the Supreme Court—all got stronger and stronger. The governments closest to the people—the state and local governments—began to depend on the federal government more and more.

Besides, that big government does many good things. It makes sure our food is safe and pure. It regulates banks so people can leave their money and not worry. It enforces laws to see that work-

In the 20th century America became an urban nation—a nation of cities. Cities don't have lakes and creeks, so these children in 1930s Harlem, New York, cooled off the city way: under the sprinkler.

Monticello (above) is the only residence in the U.S. on the United Nations' World Heritage list. (The Taj Mahal is on the same list. Do you know what and where that is?) If you can, plan to visit Monticello—you can tour its rooms and gardens. And, while you are in the area, check out Mr. Jefferson's university, too.

ing people get fair wages and fair opportunities. The federal government even provides money for hot lunches for schoolchildren.

Some experts are trying to explain that to Jefferson, Madison, and Washington. After all, this is a modern world. The United States is huge; it has many citizens; it has enormous cities. The Founders' ideas were fine in the 18th century, and the 19th too, but things are different today. Wait a minute—some of those experts are arguing. Not everyone agrees. Watch out. The experts may start fighting.

Jefferson and Madison are looking better. They are getting very interested in the 20th century, although they are a bit homesick for their own times. Well, we can jet them home—to Philadelphia, Mount Vernon, Monticello, and Montpelier. That makes them feel just fine. Their houses are still there, much as they left them.

Back in the nation's capital, the Founding Fathers are curious about all the well-dressed people on the streets. A guide tells them that these people have come from places like China, Cuba, Ethiopia, Russia, Pakistan, and Chile—and that all are now American citizens. "What a country!" says George Washington. "What a century!" says Tom Jefferson. "A fascinating time," says Jemmy Madison.

"But how about our ideas and our constitution? Can a political order fashioned for a nation of independent farmers work in a world of cities and high technology?" Jefferson is asking Ben Franklin. "That," says the old philosopher, "is the question this society must answer for itself."

1 War's End

Calamity Jane was one of the huge heavy-artillery guns that fired a final fusillade at 10:59 A.M. on November 11, 1918.

In Europe, in 1918, on the 11th hour of the 11th day of the 11th month, it suddenly became quiet. The cannons were still. For the first time since 1914, men could hear each other without shouting. The Great War—soon to be known as World War I—was over.

It had been a horrible war. Nine million men died. It was not fought soldier against soldier, like medieval battles of knights in armor. The new weapons of killing—machine guns, tanks, long-range artillery, grenades, and poison gas—led to mass slaughter. "War," wrote one soldier, "is nothing but murder."

But now the guns were silent; the dying was finished.

In Washington, D.C., even though it was six o'clock in the morning, America's 28th president, Woodrow Wilson, was up and at his desk. Because he was considerate, and feared his clackety typewriter would wake his wife and staff, he sat and wrote these words with a pen on White House stationery:

> *Everything for which America has fought has been accomplished. It will now be our fortunate duty to assist by example, by sober, friendly counsel, and by material aid, in the establishment of just democracy throughout the world.*

They were the words of a high-minded leader. The slim, frail, bookish man had proved to be a great war president. In amazingly fast order he had turned a peaceful nation into a strong fighting

Woodrow Wilson

Stretcher bearers carry the wounded from the ruined French town of Vaux, which was captured by the U.S. Army's 2nd Division.

No one could buy his way out of service in World War I (unlike the Civil War). And for the first time women served officially in the armed forces.

13

Left, officers of the 129th Field Artillery. Second row, third from right, is a captain named Harry S. Truman. (More about him at the end of this book.) Right, Company M, 6th Regiment, greets the Armistice.

force. The country's factories had gone from making corsets, bicycles, and brooms to production of guns, ships, and uniforms. In just over a year—beginning in April 1917—more than a million American men had been drafted into the army, trained, and sent overseas. And just in time. In Europe the fighting had been going on for three years; both sides were near collapse.

It had been a heartbreaker of a war—awful, dreary, bloody—begun in Europe for selfish reasons. It ended up making nations and people cruel, and bitter, and angry, and it led to another terrible war.

The Central Powers (Germany, Austria-Hungary, and the Turkish Ottoman Empire) were on one side, against the Allies (Britain, France, Russia, Japan, and Italy), with a few other nations involved, too.

The Germans had taken a gamble. Before the United States entered the war, Germany sank neutral American ships carrying food and supplies. American lives were lost. The Germans knew that might bring the United States into the war. They

Gee, How They Sang!

Lieutenant Harry G. Rennagel of the 101st Infantry wrote his family:

Nothing quite so electrical in effect as the sudden stop that came at 11 A.M. has ever occurred to me. It was 10:60 precisely and—the roar stopped like a motor car hitting a wall. The resulting quiet was uncanny in comparison. From somewhere far below ground, Germans began to appear. They clambered to parapets and began to shout wildly. They threw their rifles, hats, bandoliers, bayonets, and trench knives toward us. They began to sing. Came one be-whiskered Hun with a concertina and he began goose stepping along the parapet followed in close file by fifty others—all goose stepping....We kept the boys under restraint as long as we could. Finally the strain was too great. A big Yank named Carter ran out into No Man's Land and planted the Stars and Stripes on a signal pole in the lip of a shell hole. Keasby, a bugler, got out in front and began playing "The Star-Spangled Banner" on a German trumpet he'd found in Thiaucourt. And they sang—Gee, how they sang!

weren't worried. They thought it would take several years for the United States to get ready to fight. By that time they expected the war in Europe to be over. Most German leaders believed that the American system of government was very slow.

The scholarly, honorable man who was president stunned them. He was stronger than they thought possible. He asked Congress for special war powers; he was able to act quickly.

Woodrow Wilson's greatest strength was his integrity. People trusted him because they knew he was trustworthy. He inspired others. He believed in the American dream—in Jefferson's words about how all people have a right to "life, liberty and the pursuit of happiness." Wilson wanted to see that dream spread around the world. He convinced the people of the United States to go to war without thought of gain for themselves. He made it clear to everyone that America's only goal was "to make the world safe for democracy." He made America's participation in the war seem noble and unselfish.

It was still dark, but on November 11, 1918, the news of war's end was too good to wait for daybreak. Whistles tooted, church bells rang, and sirens blared. Before long the streets across the nation were filled with people cheering, shouting, hugging, and kissing. America had gone to war and the world was going to be a better place because of it, or so it seemed on that Armistice Day.

The morning is chilly. Mother and I wear sweaters as I follow her around the big old house. Suddenly bells begin to ring, the bells of Yamhill [Oregon]'s three churches, and the fire bell. Mother seizes my hand and begins to run, out of the house, down the steps, across the muddy barnyard toward the barn where my father is working. My short legs cannot keep up. I trip, stumble, and fall, tearing holes in the knees of my long brown cotton stockings, skinning my knees.

"You must never, never forget this day as long as you live," Mother tells me as Father comes running out of the barn to meet us.

Years later, I asked Mother what was so important about that day when all the bells in Yamhill rang, the day I was never to forget. She looked at me in astonishment and said, "Why, that was the end of the First World War." I was two years old at the time.

—BEVERLY CLEARY,
A GIRL FROM YAMHILL

The New York Times.

ARMISTICE SIGNED, END OF THE WAR! BERLIN SEIZED BY REVOLUTIONISTS; NEW CHANCELLOR BEGS FOR ORDER; OUSTED KAISER FLEES TO HOLLAND

Armistice Day, New York City. "The...crowds," said one observer, "rarely raised a cheer....It was enough to walk...with 10,000 strangers, and to realize in that moment of good news not one of them was really a stranger."

2 Fourteen Points

President Wilson in London with King George V. Wilson stayed in Buckingham Palace, which was freezing (due to wartime coal shortages). The king gave him a small electric heater—it didn't help much.

The innocent, optimistic, sure-of-itself 19th century didn't actually end in America until the First World War began. The real start of the 20th century came in 1917. No question about it, the war changed things. It changed people. They began to question old ideas that had never been questioned before. Hardly anyone seemed sure of anything.

Except Woodrow Wilson. He was like an old-time Puritan, convinced of God's grace and very sure of himself. Wilson would do everything possible to lead his nation and the world on a path of righteousness. His father had been a minister; he had the preacher's genes. He spoke eloquently and told the world how to behave. Unfortunately, some people don't like being told what to do—even if the teller is right.

Before the war ended Woodrow Wilson came up with "Fourteen Points" on which the peace was to be based. Wilson didn't believe in revenge; he believed in the power of kindness. He said he wanted "peace without victory." Now that was a startling statement in a nation that had cheered Ulysses Grant when he called for "unconditional surrender." But Woodrow Wilson had grown up in the defeated South. He knew about the hatreds that can come after a war. He didn't think an enemy needed to be shamed, or made poor. He intended to lead the world toward a generous and lasting peace.

Wilson's Fourteen Points may have been the most forgiving peace plan ever. Under the Fourteen Points, people all over the world were to

16

determine their own fate—by vote. It was called *self-determination*. Self-determination was to end the old imperialist system that let winning nations grasp foreign colonies. The Fourteen Points also called for:

- *free trade (that means no tariffs)*
- *an end to secret pacts between nations*
- *freedom of the seas*
- *arms reduction*
- *the forming of a world organization—a League of Nations*

Wilson expected that league to guarantee freedom to all the world's peoples and keep the peace between nations.

Leaflets describing the Fourteen Points were dropped over Germany from those new vehicles that had been used, for the first time, as instruments of war: wood-framed airplanes. The German people—who were tired of the war and close to rebellion—read the leaflets, hoped for peace, and soon forced their ruler, the Kaiser (KY-zer), to flee the country.

With the war over, Wilson set off for Europe, the first American president ever to do so while in office. He wanted America to lead the world to a just peace, and he wanted to be the peacemaker. The European people were wild with admiration for Woodrow Wilson. They greeted him with flowers and cheers. They called him the savior of the world.

Too bad he went, say some historians. Others say it would have been worse if he'd stayed at home. Everyone agrees: Wilson didn't get what he wanted. Perhaps because of that, the Great War, which was called the "war to end wars," didn't end anything. It turned out to be World War I. Another world war—which was much worse—followed 21 years later.

What went wrong? Why didn't Wilson get his just peace?

Was it because he was too sure of himself? Or because he didn't worry enough about jealous politicians, at home and in Europe? Was it the tragedy of his health?

Wilson is fêted in Europe when he first arrives. But despite his efforts, the Germans do not get a generous peace. Later, they will say they were "stabbed in the back." Their sense of betrayal will have awful consequences.

The Big Four at Versailles: left to right, Lloyd George of Britain, Orlando of Italy, Clemenceau of France, and Wilson. "England and France," wrote Wilson, "have not the same views with regard to the peace that we have by any means."

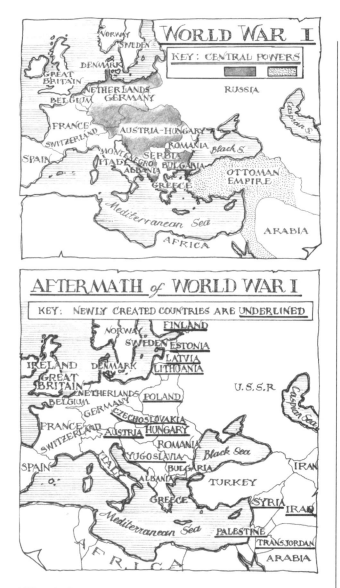

Wilson's Fourteen Points provided for self-determination of the peoples of Europe. But in fact many of the new national boundaries were decided by the Allied politicians in secret meetings where they drew lines and argued over huge maps.

Premier (which is French for *first*) is often used to mean *prime minister* or *president*.

(Before he left the presidency, he exhausted himself, lost contact with reality, and became unable to fight for his beliefs.) Maybe it was all of those things—and more, too. After four years of war, many Americans seemed to have stopped caring. Most just wanted to get on with their lives; some didn't want to be bothered by ideals; others were disappointed that we hadn't smashed the enemy. Besides, President Wilson's sermons were getting tiresome.

France's crafty old premier, Georges Clemenceau (cleh-mon-SO)—who was called "the Tiger"—said, "God gave us his Ten Commandments and we broke them. Wilson gave us his Fourteen Points—we shall see."

What Clemenceau saw was that France did, indeed, want revenge. Germany had invaded France twice within his memory (in 1870 and 1914). Two generations of young Frenchmen were dead. The French countryside was devastated. The French wanted protection and repayment for what they had suffered. They, and England and Italy, wanted—and got—a hard peace. They were angry with Germany.

The peace treaty was signed at a gorgeous French royal palace called Versailles (vair-SY). Some of Wilson's most important points got thrown out the windows at Versailles. Germany was blamed for the whole war and given a huge bill for war costs. The Germans (who had surrendered, in part, because of their faith in the Fourteen Points) felt betrayed. But the idea that meant most to Wilson—the League of Nations—was saved. He believed that the League would right the wrongs of the Old World order.

And it might have done so, if the nation that was now the most important power in the world had joined the League. (What nation could that be?)

American treaties with foreign powers must

be agreed to by two-thirds of the members of the Senate—a simple majority won't do. At first, most Americans believed in the League of Nations. But there were strong senators who hated Wilson. Some were Republicans who were anxious to win the next election; they thought that a triumph for Wilson would hurt their party's chances.

When Wilson went to Europe he brought many advisers with him; they were either professors or Democrats. None were prominent Republicans. That wasn't wise or generous on Wilson's part. Some Republican senators began to fight the idea of the League. Many Americans, Democrats as well as Republicans, worried about America getting involved in Europe's problems.

Woodrow Wilson knew that the problems of one part of the globe were now the problems of all peoples. America could not hide from world responsibility. So the president decided to do what he did best: explain things to the American people. That had worked for him before. But, in those days before radio and TV, it meant getting on a train and giving speeches. Wilson crossed the country; he gave three or four speeches a day talking about the importance of the League of Nations.

Wilson wanted "open covenants of peace, openly arrived at." He did not get them at Versailles (above).

Wilson underestimated Republican opposition to the League of Nations—and ordinary Americans' lack of interest in it. With Congress against him, he took his treaty to the people. But he failed to drum up enough enthusiasm.

19

Senator Henry Cabot Lodge of Massachusetts led Republican opposition to the League.

It was too much for his health. Wilson had been working hard. In Paris he had been ill and had acted strangely. In Pueblo, Colorado, he was so sick he could not finish his speech. Then he had a stroke. He was never the same again.

Those who opposed the League in the Senate were now able to defeat it. The United States did not join the League of Nations. You can imagine how Woodrow Wilson felt. He believed that without a strong league to enforce peace, there might be another war—and that it would be much worse than the Great War. "What the Germans used were toys compared to what would be used in the next war," he said.

But we didn't listen. The United States embarked on a period of "isolation." We tried to stay away from the rest of the world and its concerns. We would learn that could no longer be done. Like it or not, the United States was now a world leader.

Justice Oliver Wendell Holmes, Jr.

In 1917 Congress passed an Espionage Act. In 1918 it passed a Sedition Act. (*Espionage* is spying; *sedition* means inciting others to rebel.) Those acts were meant to ban speech that might harm the war effort.

The First Amendment guarantees free speech. Were these acts unconstitutional? Or do things change in wartime? Clearly, war demands national unity.

No Clear and Present Danger

When some anarchists threw 5,000 anti-war leaflets from a New York hat factory, and were arrested and convicted, the case was appealed all the way to the Supreme Court. (The pamphlets called for a strike by weapons makers.)

The court upheld (agreed with) the convictions, but two justices—Louis D. Brandeis and Oliver Wendell Holmes, Jr.—disagreed. Justice Holmes's dissent has become more famous and more often cited than the majority opinion. He said that speech may be punished only if it presents "a clear and present danger" of producing evils that the Constitution tries to prevent. "Now nobody can suppose that the surreptitious publishing of a silly leaflet...would present any immediate danger." His opinion was that "the defendants were deprived of their rights under the Constitution of the United States." Today, the concept of *clear and present danger* is used as a test of whether speech should be censored. Holmes became known as the "Great Dissenter" for this and other strong opinions that were contrary to the majority of the court.

The Supreme Court invites Brandeis to join her ranks while fat cats look on in horror.

3 Another Kind of War

INFLUENZA
FREQUENTLY COMPLICATED WITH
PNEUMONIA
IS PREVALENT AT THIS TIME THROUGHOUT AMERICA.
THIS THEATRE IS CO-OPERATING WITH THE DEPARTMENT OF HEALTH.
YOU MUST DO THE SAME
IF YOU HAVE A COLD AND ARE COUGHING AND
SNEEZING DO NOT ENTER THIS THEATRE
GO HOME AND GO TO BED UNTIL YOU ARE WELL
Coughing, Sneezing or Spitting Will Not Be
Permitted In The Theatre. In case you
must cough or Sneeze do so in your own hand
kerchief and if the Coughing or Sneezing
Persists Leave The Theatre At Once.

This Theatre has agreed to co-operate with
the Department Of Health in disseminating
the truth about Influenza, and thus serve
a great educational purpose.

**HELP US TO KEEP CHICAGO THE
HEALTHIEST CITY IN THE WORLD**
JOHN DILL ROBERTSON
COMMISSIONER OF HEALTH

The usual October death rate from influenza and pneumonia was 4,000. In 1918 it was about 194,000.

I had a little bird,
And his name was Enza;
I opened the window,
And in flew Enza.

In flew Enza—say it fast and it becomes "influenza." It was a catchy little rhyme, and boys and girls skipped rope to it. It was also an epidemic; no, it was worse than that. It was a *pandemic*, which means a disease that spreads across many nations. This one went around the globe. And it was deadly.

The word ***influenza*** first appeared in 1743 after an epidemic in Italy. It is an Italian word, related to *influence*, and it means an "intangible visitation" (a visit by something you can't touch). ***Pandemic*** comes from the Greek words *pan* ("all") and *demos* ("people").

Diseases don't fly in the window, but the influenza of 1918 almost seemed to. It lasted about nine months, and, worldwide, killed 20 million people. That was more than the total of deaths during the four years of the Great War. Mysteriously, it struck at about the same time in India, and Russia, and China—no major nation escaped. In the United States there were more than half a million victims. On one terrible day in Philadelphia, almost 1,000 people died. Neither doctors, nor hospitals, nor cemeteries could handle the awful burdens put

Returning soldiers, and anybody who didn't wear a mask, could be fined $100 and jailed. *Obey the laws, and wear the gauze, protect your jaws from septic paws*, went one ditty. But the masks were useless.

A public-health doctor in Washington, D.C., found that the only way he could be sure of having enough room in his emergency hospital was to keep undertakers always waiting outside the door so that the dead could be taken away immediately.

upon them. In those days before the discovery of modern medicines, there was little the doctors could do.

In New York and Chicago, laws were passed making it illegal to sneeze or cough in public without using a handkerchief. Police dutifully hauled sneezers and coughers to court, where they were given stiff fines. The police had time to worry about influenza because the robbers and murderers were sick, too. In October 1918, Chicago's crime rate dropped almost by half.

The epidemic spread most rapidly in cities—where people are crowded together—but many in the countryside died too. A prominent senator lost a son and daughter. Soldiers, fighting heroically against enemies they could see, fell to invisible germs. In America the flu took 10 times as many lives as the war. The last week of October in 1918, 2,700 American soldiers died fighting in Europe; that same week, 21,000 Americans died at home of the flu.

It was called Spanish influenza—because people

Today, scientists believe that the 1918 influenza epidemic may have stopped killing people because of the way the influenza virus behaves. Dr. Michèle Barry, an infectious-diseases specialist at Yale University Medical School, says: "Every year, the influenza virus changes the coat of protein that surrounds it. Some protein coats seem to make the virus weaker or stronger. We think that during the course of the 1918 epidemic, the virus changed its protein coat and became weaker." She adds that it's also possible that after the disease killed off the most vulnerable victims (especially old people), the people who were left were tougher and less likely to die.

New York City's phone company begged people to make only urgent calls—many switchboard operators were hit by flu, too.

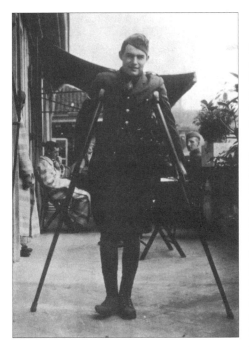

One of the wounded soldiers who came home to find things changing was a young man named Ernest Hemingway (left), who was soon writing novels and stories (of war and life between the wars).

thought it had started in Spain. It hadn't. It may have begun in the United States, from a disease of hogs (it is sometimes called *swine flu*, after the hogs). Some say it was the worst pandemic in history. It wasn't that it killed the most people, it was that it killed so rapidly. Someone figured, mathematically, that if it had continued spreading for another year, at the rate it was going, the world's population would have been wiped out.

By Armistice Day, November 11, the peak had passed. The disease soon departed as mysteriously as it had arrived. It left the country exhausted. Wasn't a war trouble enough? Everyone had worked hard supporting the war effort. Americans had done astonishing things in factories and on the farms. They'd fed Europe with an amazing harvest of grain; they'd armed the Allies. Citizens had given up luxuries and even some necessities to help others. That flu epidemic was the final straw. Someone needed to find something to cheer people up.

Soon a new word was being used. It was *normalcy*. That's what people wanted. They wanted to go back to the good old days before the war. But time won't march backward.

Those boys and girls who were skipping rope in 1918 had no idea what was ahead of them. They wouldn't have believed it if you had told them. Normalcy? No way. They were going to live in a world of radio, TV, computers, jets, and rockets. In 1918 that was the stuff of science fiction. Their world

If you were sick in 1920, you didn't go to the doctor's office. The doctor came to your house. He brought a black satchel with him. It was stuffed with medical supplies: pills, salves, bandages, and maybe a stethoscope and a thermometer. Those two pieces of equipment were about as high-tech as medicine got. The world of the modern hospital, and sophisticated equipment, was decades away.

From Art to Science

Abraham Flexner, an educator who was asked (by the Carnegie Foundation) to study medical schools in America, inspected each of the country's 155 medical schools, and, in 1910, described them in a study that became very influential. Mostly, he thought American medical schools weren't very good and that they turned out poorly trained physicians. Flexner was especially critical of women's medical colleges (there were 16 of them) and black medical schools (there were 10). Doctors then earned little money and relied chiefly on experience and observation in treating patients. Flexner said that medicine needed to be a science. His report helped make it just that. It also drove most women and minorities from the practice.

was slow-paced, and mostly powered by horses and mules.

Their older brothers—the soldiers who came home from Europe in 1919—had exciting things to tell them. They'd been to Paris and had seen fancy night clubs, stunning buildings, and splendid boulevards. Some bragged about their heroism in battle, which was understandable; you had to be tough—or lucky—to be a survivor. A few came home without arms or legs. Some didn't want to talk about the war at all. They, too, were looking for normalcy.

The returning soldiers were surprised to find that America had changed in the year they'd been gone. They noticed two things right away. One had to do with beer and liquor. During the war it was considered unpatriotic to drink alcohol. Beer is made from grain, and grain was needed to feed soldiers. Now that the war was over, many people wanted to put an end to all liquor drinking. It would make the world a much better place, they said.

The other change gave some of the soldiers a chuckle. Imagine, women were demanding equal rights: they wanted to be full citizens. Why, soon they'd probably want to wear pants, too!

The Wild Beasts

It was a cold winter day in 1913 when the doors opened at an old, drafty armory on 25th Street and Lexington Avenue in New York City, and people got to see some new paintings and sculpture (including many from Europe). American art was never the same again. Some of the paintings in the Armory Show were done by Henri Matisse, Paul Cézanne, Pablo Picasso, Paul Gauguin, and Vincent van Gogh. Today we know them all as great artists, but in 1913 their work was unlike anything most people here had seen before. Matisse was part of a French group called Les Fauves, or "the wild beasts." Many viewers thought it a good title for all the artists.

One painting, by Marcel Duchamp, was called *Nude Descending a Staircase*. Classical paintings often showed naked figures reclining on couches. This painting showed what seemed to be a bunch of sticks—or maybe a figure; it was hard to tell—but something very active *was* happening on the canvas. It was worth a second, and even a third look. Many of the paintings included unrecognizable objects. Before, art had always more or less imitated reality. These paintings were completely different. In Chicago some art students burned effigies (stuffed figures) of Henri Matisse. But many Americans were profoundly changed by the new modernism. The artist Stuart Davis said that the Armory Show was "the greatest single experience…in all my work." He was not alone.

Marcel Duchamp's Nude Descending a Staircase; *to some it was an outrage.*

4 The Prohibition Amendment

You couldn't get drier than a camel—it became the symbol of Prohibition party supporters.

The Constitution does not give Congress the right to tell people what they may eat or drink. If someone wants to drink poison, only a state can make laws to try to keep him from doing so.

Many people say that alcohol can be a kind of poison. No one disagrees that drinking too much is harmful.

Drunkenness was a special problem in early America. Most drinking was done in saloons, where women were not admitted. Some men took their paychecks,

The ladies of the Women's Christian Temperance Union campaigned "for God, for home, for native land." But the lawmakers found that it was hard to make a crime out of drinking, which many had never seen as a crime.

Prohibition agents got to work disposing of booze. But there were only 1,500 agents, not really enough to enforce the law—especially when they were up against the ruthless gangsters who sold the liquor.

went to a saloon, got drunk, and then went home drunk, with no money left for their families. Reformers decided to attack the problem. Some of them believed in *temperance*, which means moderation. Others believed in *prohibition*, which means outlawing all drinking.

Some women's groups fought for prohibition. Several religious groups—especially Methodists and Baptists—joined the battle. Many states became *dry*. In a dry state it was against state law to buy or sell liquor. Some people wanted to go further. They wanted the whole nation to be dry. A

There were more ways of hiding alcohol than strapping it to your leg. The man around the corner is pouring his stash into a hollow cane.

constitutional amendment was needed.

It was the Progressive Era: people thought that laws could help make people perfect—or close to it. It took about 20 years to get the 18th Amendment passed, but finally it was done. The Prohibition Amendment became law in 1920. The amendment made it illegal to sell liquor anywhere in the United States. Most people thought it a very good idea. All but two states passed the Prohibition amendment.

It didn't work. Many people who wanted to drink kept drinking (although

The Constitution of the United States, Article V

The Congress, whenever two thirds of both Houses shall deem it necessary, shall propose Amendments to this Constitution, or, on the Application of the Legislatures of two thirds of the several States, shall call a Convention for proposing Amendments, which, in either Case, shall be valid to all Intents and Purposes, as Part of this Constitution, when ratified by the Legislatures of three fourths of the several States, or by Conventions in three fourths thereof.

What does all that mean? Read it slowly and it isn't as difficult as it may seem. What it means is that the men who wrote the Constitution—James Madison, Gouverneur Morris, Ben Franklin, John Adams, and the others—understood that a constitution needs to be adaptable. The Founding Fathers wanted people in the future—you and me—to be able to change the Constitution. But they didn't want to make it too easy to change. If they did that the Constitution wouldn't have much lasting value: it would get changed all the time.

So they came up with the idea of amendments as a way to change the Constitution. It has been more than 200 years since the Constitution was written; hundreds of amendments have been proposed, but only 26 have been passed.

For an amendment to succeed, two-thirds of the Congress must pass it—that means two-thirds of both the Senate and the House of Representatives. Then three-fourths of the states must also approve the amendment. (The Constitution may also be amended if a constitutional convention is called by three-fourths of the states—that has never been done.)

Remember, we have a federal form of government. Power is shared among the national government in Washington, D.C., and the state governments. The Constitution lists all the things the president, Congress, and the courts can control. Any powers not listed in the Constitution belong to the states.

Per capita (pur-CAP-it-uh) is Latin, and means "by the head" or per person. In other words, the total amount of liquor consumed in the U.S., divided by the number of people in the U.S., showed there was less alcohol drunk during Prohibition than before. But many new kinds of people began drinking; that was the problem.

After the Prohibition amendment was passed, Congress needed to provide for its enforcement. That was done with a law called the Volstead Act. Prohibition didn't make it illegal to drink, or even to buy liquor; it just made it illegal to sell it.

per capita alcohol consumption did fall during Prohibition years).

But some people, especially some women and young people, who had not drunk before, decided to try it. Prohibition wasn't supposed to do this, but in some crowds it made drinking fashionable. (Maybe it had to do with disillusionment after the war. Writers were calling this a "lost generation." People weren't really lost, but they were confused about right and wrong.)

Since selling liquor was now a crime, gangsters took over that activity. People who sold liquor were called *bootleggers*. (Some of them stuck flasks inside high boots.) Ships running whiskey from foreign suppliers to coastal ports were called *rumrunners*. Illegal bars, where drinks were sold, were called *speakeasies*. (If people spoke loudly, and the police heard them, the bar would be raided. So they spoke "easy.")

No one expected it, but Prohibition made crime a big business in the United States. Americans learned that some kinds of prohibition must be done by persuasion and education. Laws and force don't always work.

Another amendment was needed to get rid of the Prohibition amendment. The 21st Amendment was passed in December 1933. It ended what was a well-meaning experiment. The experiment had failed.

But how do you get people to stop doing something that isn't good for them? Do the lessons of Prohibition apply to drugs? Some people say we should make it legal to buy drugs; then criminals could not earn big money selling drugs. Others say that would encourage people to use drugs. What do you think?

In 1933, the country had more than 200,000 illegal speakeasies. This painting (and those on the cover and on pages 25, 27, and 35) was done by the artist Ben Shahn, who put many social and political issues of his time on canvas.

5 Mom, Did You Vote?

The suffragists had become experts at getting publicity for the cause.

We're heading back in time—just a bit. It is 1917, and some women are marching in front of the White House. They carry a big banner that says 20 MILLION AMERICAN WOMEN ARE NOT SELF-GOVERNED. In Europe, American soldiers are fighting for democracy; these women feel they should fight for it at home.

Day after day, for months, the women march in front of the president's house. They are peaceful and respectful, but persistent. Some people don't like it; they say the suffragists shouldn't annoy the president during wartime. The police tell the women to leave. "Has the law been changed?" asks Alice Paul, leader of the group. "No," says the police officer, "but you must stop."

"We have consulted our lawyers," says Alice Paul. "We have a legal right to picket."

The next day two women, Lucy Burns and Katherine Morey, are arrested. Other arrests soon follow.

On the Fourth of July a congressman speaks to a large crowd gathered behind the White House. "Governments derive their just powers from the consent of the governed," he says. Police keep the crowd orderly and protect the congressman's right to free speech.

Alice Paul went to England and learned tough tactics—chaining oneself to railings, disrupting public meetings—from British suffragists.

This is what we are doing with our banners before the White House, petitioning the most powerful representative of the government, the President of the United States, for a redress of grievances; we are asking him to use his great power to secure the passage of the national suffrage amendment.
—ANNE MARTIN

Dr. W. W. Parker of Richmond, Virginia, wrote an essay that he read to the Medical Society of Virginia. Women, he said, were "superior morally, inferior mentally, to man—not qualified for medicine or law." Then he continued, "God having finished this splendid world, placed at its grand arched gateway imperial man, stately and stalwart, with will and wisdom stamped upon his lofty brow."

At one demonstration outside the White House, men from the crowd tore the suffragists' banners down and pelted the women with eggs, tomatoes, and apples. Twenty-two banners and 14 party flags were destroyed.

Many Americans— men and women—don't bother to be active citizens. Many don't vote. The whole point of a democracy is that it gives everyone power. Those who don't vote give up their power.

In front of the White House a group of 13 women silently holds a banner with those very same words from the Declaration of Independence. Some are young women, some white-haired grandmothers; all are arrested. The women are taken to court and fined. They refuse to pay their fines—to do so would mean to admit they are guilty. They do not believe themselves guilty of any crime. The police take them to jail. More women are arrested. Anne Martin speaks out in court:

> As long as the government and the representatives of the government prefer to send women to jail on petty and technical charges we will go to jail. Persecution has always advanced the cause of justice. The right of American women to work for democracy must be maintained.

More women go to jail. They are separated from each other. Prison conditions are awful. For 17 days Ada Davenport Kendall is given nothing to eat but bread and water. Some women are held in solitary confinement. Some, who go on hunger strikes, are held down and fed against their will. Anne Martin, Lucy Burns, and Elizabeth McShane are force-fed. Burns is bruised on her lips and face; McShane throws up. Now the women have become interested in prison reform, as well as in women's suffrage. One woman writes that it is "necessary to make a stand for the ordinary rights of human beings for all the inmates."

Utter Nonsense

When Susan B. Anthony was a girl she stayed after school one day to ask the schoolmaster if she could learn long division with the boys. "Nonsense! Utter nonsense!" he told her. "A girl needs to know how to read her Bible and count her egg money, nothing more." But Susan Anthony was determined. So, slyly, she sat on a bench behind the boys and listened and worked problems and learned long division. Anthony never did what others did, unless she thought it right.

In the White House, Woodrow Wilson has other concerns. He is fighting a war—that war for democracy. Wilson says he isn't against women's suffrage—in fact he is for it—but, like many men, he thinks that most other issues are more important.

The women keep marching. All kinds of women. Rich and poor. Could it be that they understand democracy in a way the president doesn't?

Mrs. John Rogers, Jr., is arrested. She is a descendant of Roger Sherman (a signer of the Declaration of Independence). Like her plain-speaking ancestor, Mrs. Rogers says what she thinks. She tells the judge:

We are not guilty of any offense....we know full well that we stand here because the president of the United States refuses to give liberty to

Suffragists parade through New York City.

The right of citizens of the United States to vote shall not be denied or abridged by the United States or by any state on account of sex.

—19TH AMENDMENT
TO THE CONSTITUTION

31

Mass Meeting
TONIGHT
Ryman Auditorium
8 O'CLOCK
TO SAVE THE SOUTH
FROM THE SUSAN B. ANTHONY AMENDMENT
AND FEDERAL SUFFRAGE FORCE BILLS

Senator Oscar W. Underwood, of Alabama, and Ex-Gov. Ruffin G. Pleasant,
of Louisiana, Have Been Invited to Speak

Many people—women as well as men—fought desperately to block the 19th Amendment. At right, an antisuffrage propaganda poster implied that "emancipated" women abandoned their suffering families to pursue the vote.

Jeanette Rankin

Rebecca Latimer Felton

Montana's Jeanette Rankin was the first woman elected to Congress. She served two terms in the House (1917–1919; 1941–1943). Rebecca Latimer Felton, age 87, was the first woman in the Senate. She didn't do much. After a Georgia senator died, she was appointed to fill the vacancy for two days in November 1922.

American women. We believe, your honor, that the wrong persons are before the bar in this court....We believe the president is the guilty one and that we are innocent.

Now, isn't that what America is all about? The right of every citizen to speak out—even against the president.

Mrs. Rogers's cause is just, but her comments aren't quite fair. It is Congress that is holding things up, not Woodrow Wilson. But the president hasn't helped. Finally, he does. He urges Congress to pass the 19th Amendment. It is known as the Susan B. Anthony amendment. This battle for women's suffrage is not something new. Susan Anthony and her friend Elizabeth Cady Stanton began the fight in the mid-19th century. They spent their lives fighting for women's rights. So did Carrie Chapman Catt, the head of the National American Woman Suffrage Association.

Many men and women have worked hard for this cause. Most are people you have never even heard

Female Takeover

In 1920, the women of Yoncalla, Oregon, got together and made plans to take over the town government. They didn't tell anyone, not even their brothers or husbands. Men outnumbered women two to one in this community of fewer than 350 persons, but the women all voted. According to *Literary Digest*, they were "stirred by the alleged inefficiency of the municipal officials, and swept every masculine office-holder out of his job." When they went to the polls they elected an all-women's slate of town officials. Mrs. Mary Burt became the new mayor. The out-of-a-job former mayor, a Mr. Laswell, was said to be "much surprised."

The perpetrators of Yoncalla's conspiracy of women after their takeover of the town council, with Mayor Mary Burt in the middle.

about. (See if you can find a history of the women's suffrage movement in your community.) In Tennessee, Harry Burn was 24 and the youngest representative in the legislature when he got a letter from his mother. "Don't forget to be a good boy," wrote his mother, "and help Mrs. Catt put the 'Rat' in ratification."

The Tennessee legislators were trying to decide whether to approve the 19th Amendment or not. Half were for women's suffrage, half were not. Burn held the deciding vote. He followed his mother's advice. It was 1919, and Tennessee was the last state needed to ratify. The next year, 1920, America's women finally went to the polls.

Carrie Chapman Catt campaigned all over Tennessee to get the amendment ratified. "The summer heat was endless, and many legislators lived in remote villages," she said. "Yet the women trailed these legislators, by train, by motor, by wagons, on foot.... no woman faltered."

> *Said Mr. Jones in Nineteen-Ten:*
> *"Women, subject yourselves to men."*
> *Nineteen-Eleven heard him quote:*
> *"They rule the world without the vote."*
> *By Nineteen-Twelve, he would submit*
> *"When all the women wanted it."*
> *By Nineteen-Thirteen, looking glum,*
> *He said that it was bound to come.*
> *This year I heard him say with pride:*
> *"No reasons on the other side!"*
> *By Nineteen-Fifteen, he'll insist*
> *He's always been a suffragist.*
> *And what is really stranger, too,*
> *He'll think that what he says is true.*
> —ALICE DUERR MILLER, "EVOLUTION,"
> IN *ARE WOMEN PEOPLE? A BOOK OF RHYMES FOR SUFFRAGE TIMES*, 1915

6 Red Scare

"Whose country is it anyway?"
Uncle Sam takes care of "reds."

Communists were called **reds** after the red flag of the International, which was the worldwide communist organization.

Some people in America were scared by Russia's ideas. They were afraid of *communism*. Others were attracted to those ideas. Under communism, most property and goods belong to the state. People are expected to share. That sounds noble; it just never seems to work unless forced upon people. Communist nations have not been free nations.

After the world war, some people were scared that communists wanted to take over in the United States. There were a few communists in this country—but the communists were not successful.

Russia Revolts

Russia fought with the Allies in World War I until the Russian people decided they'd had enough of the war. It was more important, as far as they were concerned, to solve their own problems. They wanted to get rid of their ruler—the tsar (ZAR). They wanted to end the big gap between rich and poor in Russia. They wanted what Americans had wanted in 1776. They wanted freedom. So they had a freedom revolution.

At first, it looked as if they might get freedom. The people who overthrew the tsar (in 1917) were trying to create a democratic government. Then a revolutionary named Vladimir Lenin, who was living in Europe in exile, came back to Russia. That man changed the fate of Russia and the world. He became dictator of Russia. He didn't believe in democracy.

Things had been bad in Russia when the tsar was ruler. They got much worse under Lenin and the ruler who followed, Joseph Stalin. Lenin and Stalin brought totalitarianism to Russia. They brought repression, murder, state control, and misery. They brought an economic system called communism.

Lenin took Russia out of the war. That let Germany move troops from eastern Europe to France. It made the Great War tougher for the Allies.

What does all this have to do with U.S. history? A lot. You see, the world had become smaller. Not smaller in size, but in accessibility. At the beginning of the 19th century, it took at least two years for a ship to go from Salem, Massachusetts, to China and back. Now, with the telephone, communication was almost instantaneous. Modern technology meant that the ideas of one nation could spread quickly to others.

In September 1920, a bomb exploded on Wall Street, killing 38 people and injuring more. It fueled the fears of those who thought communists threatened the nation's existence.

Most American people were not attracted to communism.

In that same postwar time, there were also some *anarchists* in America. Anarchists don't believe in government at all. You don't have to be very smart to realize that anarchy doesn't work. But, when the anarchists looked around and saw poverty, war, and evil, they thought that this was the fault of governments. Some may have really believed that the answer was to do away with all governments. A few tried to do that by setting off bombs intended to kill government leaders. That, of course, was criminal behavior. Newspapers made big headlines of the bombs. Many Americans were frightened. But what A. Mitchell Palmer, President Wilson's attorney general, did was irresponsible and criminal. (He got away with it—but not in the history books.)

A. Mitchell Palmer

Palmer went on a witch hunt. The witches he went after were communists and anarchists. He took the law in his hands, and, in two days of raids in major cities (in 1920), agents invaded homes, clubs, union halls, pool halls, and coffee shops, rounding up nearly 5,000 people, who were held in jail, not allowed to call anyone, and treated terribly. Those without citizenship papers

Sacco and Vanzetti

Nicola Sacco and Bartolomeo Vanzetti were accused of murdering a paymaster and his guard at a shoe factory in South Braintree, Massachusetts. Did they do it? Even today, no one is quite sure. But they were convicted and sent to their deaths. Sacco and Vanzetti were anarchists, and many said it was radical beliefs that were on trial. The trial was a *cause célèbre* (which, in French, means a famous happening).

In 1789, Congress passed an Alien law. It kept certain people from emigrating to the United States. A sedition law made it a crime to speak against the government. People were jailed for their ideas. The people who supported those laws said they wanted to keep "dangerous foreigners" out of the country. At the time, the foreigners they feared were French.

Behind the red scare was a fear of foreigners. These men being taken to prison are all immigrant aliens.

were sent out of the country—to Russia. Most weren't guilty of anything.

Communists are sometimes called *reds*. Mitchell Palmer took advantage of America's fear of communism. He helped create a "red scare." He hoped it would make him president. During the red scare, Americans were not free to speak out about communism. They weren't free to criticize the government. Some people's lives were ruined.

Witch-hunting turns up every once in a while in American history. (It happened at Salem, Massachusetts, in colonial days; it happened after World War II with a senator named Joe McCarthy.) The good thing is that it never seems to last long. Persecution for ideas is not the American way.

Everywhere reds were under the bed—or, as in this cartoon, slithering under cover of the Stars and Stripes.

The First Amendment (part of our Bill of Rights) says: *Congress shall make no law...abridging the freedom of speech.* Does that mean that communists and anarchists are free to speak out here—as long as they do not engage in criminal activity or plot to overthrow the government?

Thomas Jefferson wrote: *Truth is great and will prevail if left to herself,* and *errors cease to be dangerous when it is permitted freely to contradict them.* He believed that when everyone's ideas are heard, people will make wise choices. Do you agree with him?

The Ku Klux Klan grew hugely in the 1920s. The Klan no longer limited its hatred and bigotry to blacks; it was anti-foreign, anti-communist, anti-Catholic, anti-Jewish.

7 Soft-Hearted Harding

Harding in 1882, aged 16, when he graduated from college and taught school for a year.

The two presidents sat together in the elegant Pierce-Arrow touring car. The car had running boards on its sides, the presidential seal on its door, and no roof.

Both men wore tall black silk hats and fashionable coats with black velvet collars. Woodrow Wilson's face was ash white. He had always been slim; now he seemed shrunken, like a dry reed. It had taken all his energy to walk from the White House door to the automobile. Just two years earlier, Wilson had been the world's hero. Now he was ill and ignored. His country seemed to have no use for him or his ideas, and he knew it.

Sitting next to him was the president-elect. The candidate the people had chosen—enthusiastically—as 29th president: Warren Gamaliel (guh-MAY-lee-ul) Harding. If a movie director were casting a president's part, he might pick Harding. The man *looked* presidential. His hair was silver, his eyebrows black, his skin tanned bronze, his voice golden. He was handsome, well-groomed, and distinguished-looking. He was also a good-natured, pleasant man.

The Pierce-Arrow pulled up to the Capitol, where Woodrow Wilson signed his last official papers as president. The new vice president, Calvin Coolidge—a man of few words and few ideas—was sworn into office in the Senate chambers. Then Warren Harding took his place at the center of the grandstand built for his inauguration in front of the Capitol. He put his hand on the Bible and swore to execute the office of president to the best of his ability. He had chosen a line from the

A ***running board*** was a long entry step that ran along the side of an automobile.

When is an elephant like a teapot? When a Republican president's friends rob the government.

The scandal in the Harding administration centered on some naval oil reserves, including one at Teapot Dome, near Casper, Wyoming. (Geologists call a swollen upward curve in the earth's surface a dome.) Congress was concerned that the military services have enough oil in case of emergencies, so it had set aside certain oil-rich areas for government use. Harding had appointed men who gave secret rights to those government oil lands to individuals and to private companies who got rich on oil meant for public use.

Old Testament book of Micah, and he read it in a clear voice amplified by loudspeakers:

What doth the Lord require of thee, but to do justly, and to love mercy, and to walk humbly with thy God.

The Marine band played "America."

After the stormy war years, this low-key, modest inaugural seemed just right. "We must strive for normalcy," said Warren Harding, and the public applauded.

Two years and five months later, Harding was dead, and the nation wept as it had not done since the death of Abraham Lincoln. In his lifetime, Warren Harding was one of the most popular presidents ever.

He is said to have died of heart failure, but perhaps it was of a broken heart. For he knew,

The mustachioed men are Interior Secretary Albert Fall (left) and Edward Doheny. Doheny bribed Fall to lease the Teapot Dome oil fields.

Montana Senator Thomas Walsh investigated Teapot Dome. He was harassed by the FBI, which tapped his phones, opened his mail, and made anonymous threats on his life.

before he died, that his friends, whom he trusted with important government jobs, had betrayed him and the nation. They had stolen and plundered. They had given away priceless oil reserves, they had laughed at the conservationists and taken bribes from business and criminal interests. They had become very rich.

And what did the American people think when they heard the news? At first they were angry at the senators and journalists who told them. Later, they became angry when they thought about Harding. Before long, historians were calling him the worst of all presidents.

He wasn't the worst. A few others were equally inadequate. But he never should have had the job. He wasn't smart enough. It is a myth that he didn't work hard, however. He did (although he also played a lot of poker and golf). He just couldn't seem to handle the presidency. Maybe he wasn't tough enough. He wanted everyone to like him; but a president has to make hard decisions.

If you are president you need to appoint good people to the cabinet and to thousands of administrative jobs. You need to be a leader of the armed forces. You need to make a huge budget to run the country. You need to come up with ideas for domestic policies that will make the country prosperous and happy. You need to make foreign policies to guide the nation in its relations with other nations. You need to get along with Congress. You need to get along with the states and their governors. You need to be a role model for millions of citizens with differing ideas and desires. Being president is an enormous and complicated job.

Warren Harding, as I said, was a pleasant man, always gracious and considerate. Maybe it was because he was so

Loudspeakers were used at Harding's inauguration for the first time in the event's history.

"Lord, Lord, man!" said one of Harding's aides to reporter William Allen White of the *Emporia Gazette.* "You can't know what the president is going through. You see he doesn't understand it; he just doesn't know a thousand things he ought to know. And he realizes his ignorance, and he is afraid. He has no idea where to turn."

Political Advice: Don't Say Yes—or No

Harding was the first president to address the nation over the radio.

Warren Harding, 29th president, was the owner and editor of a newspaper in Marion, Ohio, when he entered politics. He became a senator. In 1920, the Republican National Convention couldn't decide on a candidate—the delegates were deadlocked—so on the 10th ballot the party turned to well-liked, easygoing Senator Harding, and nominated him. During his presidential campaign, Harding wouldn't say if he was for the League of Nations or against it. His opponent, Democrat James M. Cox (also of Ohio), said he was for the League. Cox lost the election.

good-hearted that he wanted to help his old friends. Maybe that's why he put some of those friends in important jobs. Most weren't qualified for those jobs. Some were crooks who stole a whole lot of money from the nation.

Harding didn't realize it. But he should have. That was his job. A more able president might have saved the country a lot of grief.

Under Harding, a cartoonist said, the government had put the whole country up for sale.

Chicken-Bone Specials

Between 1910 and 1920, more than one million black people headed north. Working conditions in the South were awful; schools were worse; and most blacks couldn't vote. The North held the hope of better jobs, better schooling, and a chance to get ahead. For a while, the Pennsylvania Railroad offered free passage to blacks who could recruit others to come north. The trains came to be called "chicken-bone specials"; blacks weren't allowed to eat in the dining cars, so they had to bring their own food—usually fried chicken. New York's black population (centered in Harlem) increased by 66 percent in that 10-year period. Chicago's black population (centered on the South Side) increased by 50 percent. That black migration—from field to factory, from rural to urban—continued through most of the century. Jacob Lawrence painted the migration in a series of paintings that are small in size but powerful in impact.

"There is absolutely nothing before them on the farm," said a government report on southern labor, "no prospect...but to continue until they die." Instead, they fled.

8 Silent Cal and the Roaring Twenties

You could hardly imagine Coolidge doing the Charleston in short skirt and beads. But there was no doubt that he let big business call the tune.

Thrifty, tight-lipped, humorless, Vermont-born Calvin Coolidge became president when Harding died. What did the American people think of him? The same as they had thought of Harding while he was in office. They believed him a splendid president. The times were prosperous. Coolidge was very popular. What do historians say of President Coolidge today? "A respectable mediocrity," says one. What is a mediocrity (me-de-OCK-rih-tee)? A very ordinary person.

Why the difference of opinion? Can't the American people tell a good president from an ordinary or not-so-good one? Not always, it seems.

Historians happen to have a big advantage: hindsight. They know how things come out. When you are living through a time it isn't so easy to judge it.

And the 1920s were a confusing time. They were called the "Roaring Twenties" because everyone seemed to be intent on having a good time. The decade has also been called the "Jazz Age" and the "Dance Age." It would have been fun to be alive in the '20s.

It was a time of great change. In 1919, before the '20s began their roar, women's ankles sometimes could be glimpsed beneath long skirts. Those ankles, however,

1925: A Caviar Year

It is 1925 and the '20s are roaring! Wyoming elects the first woman governor in U.S. history, Nellie Taylor Ross; *The New Yorker* is introduced as a magazine for "caviar sophisticates...not for the old lady in Dubuque." The first issue costs 15 cents. At Nome, Alaska, dog-team relays bring serum to combat a serious diphtheria epidemic. Born in 1925: Robert F. Kennedy, Rod Steiger, and Malcolm X.

The '20s were the age of the fad. Dancing was the biggest fad of all, and marathon dancing carried it to extremes. Dancers competed for prize money to see who could last longest. For some poor contestants it wasn't a game.

"Teaching an old dog new tricks"—won't you Charleston, Charleston with me?

Women wearing bathing suits were measured to see if they were showing too much leg. Here, Chicago police make an arrest for "indecent exposure" in 1922.

were modestly hidden behind high-topped shoes. Then skirts started going up, and up, and up.

That made it a tough time to be the parent of a girl. It wasn't easy to be a girl then either. Most young women were cutting their hair—short. They called it *bobbing*. Some parents wouldn't allow it. Short hair seemed indecent to the older generation, but up to the minute to those who did it. The girls who weren't allowed to cut their hair felt old-fashioned.

Some daring women were wearing bathing suits that left their legs uncovered. Police arrested women on the beaches for doing that. And makeup! "Nice" women started wearing lipstick, rouge, and powder. The older generation worried. "What is the world coming to?"

Those girls who bobbed their hair and wore short skirts and lipstick were called *flappers*. They did other things, too. They drove automobiles, got jobs, went to the movies, read romantic novels, played ping-pong, and danced. My, did they dance! It was the big thing in the '20s. And the big dance was the Charleston. (In New York City, Gimbel's department store advertised special Charleston dresses that swung loose on the body. The price was $1.58.) When you danced the Charleston you swung your arms, kicked up your heels, knocked your knees together, and moved frantically.

Frantic is a good word to describe the '20s. The idealism of the Wilson years seemed to have come to nothing. After the war, everything was supposed to be better. But anyone could see that it wasn't. And there was the Prohibition idea. Americans wanted to have a good nation—where all people behaved themselves and didn't get drunk—but that wasn't working either. If you read the newspapers you could see that criminals were becoming rich and powerful selling liquor. So maybe the best thing to do was to forget about ideals and have a good time—frantically—which was what a lot of Americans did in the '20s.

It was a materialistic age. People concentrated on making money and buying things for themselves. Successful businesspeople became national heroes. There were more rich people than ever before in American history. No one seemed to notice, however, that there were also growing numbers of unemployed people—people who were desperately poor. And many farmers were in terrible trouble.

But for most Americans, the times seemed good. The stock market—like women's hems—went up and up and up. Land values boomed. People were able to buy things they never could buy before. In 1920 the car was a novelty. Ten years later, almost every family had a car. Many Americans who didn't have indoor toilets in their homes had cars in their yards. The automobile

A Death in the Family

Calvin Coolidge had been an active governor of Massachusetts. When he became president after Warren Harding's unexpected death, it looked as if he would be an energetic and decisive chief executive. He worked long hours and voiced strong opinions. Then something terrible happened. His son, 16-year-old Calvin, Jr., died of blood poisoning, which developed from a toe blister he had gotten playing tennis. The president went into a period of deep depression. Almost no one, besides his close associates, knew that was happening. Coolidge was overwhelmed with grief. He began taking long naps. He paid little attention to his work. He developed an assortment of illnesses. He became an indifferent president.

1927: The U.S. Supreme Court in *Nixon* v. *Herndon* rules unanimously that a Texas law forbidding blacks to vote in primary elections violates the 14th Amendment and is unconstitutional.

By the time Henry Ford brought out his Model A, in 1927, there were 21 million cars in America—and the traffic jam was beginning to be a familiar phenomenon.

was becoming a necessity.

Before the war, life had been slow-paced; now change was coming with cyclone speed. Ordinary people owned radios and listened to comedy shows and the nightly news. In Florida, in 1924, a schoolboy named Red Barber heard radio for the first time at a friend's house. Barber was so excited he stayed up most of the night listening to news from around the nation. It was a new experience. "A man…in Pittsburgh said it was snowing there…someone sang in New York …a banjo plunked in Chicago…it was sleeting in New Orleans." (Red Barber later became a radio sportscaster.)

Young people were flocking to the movies and, in 1927, movies began to talk. Talk about fun!

A legend and his creator: Mickey and Walt.

The following year, in Hollywood, California, a young filmmaker named Walt Disney produced the first animated sound film, *Steamboat Willie*, and introduced a little mouse named Mickey to the American public.

Suddenly, America seemed filled with artistic geniuses: musicians George Gershwin and Aaron Copland; writers Ernest Hemingway, William Faulkner, and F. Scott Fitzgerald; and artists Mary Cassatt, Grant Wood, and Thomas Hart Benton. And those are just a few of the names.

Harlem (a part of New York City with a rapidly growing black population) began vibrating with artistry. It was contagious. Playwrights, poets, musicians, artists, and actors, all living within a few blocks of each other, were sharing ideas. Langston Hughes, Claude McKay, and Countee Cullen began writing poetry. Zora Neale Hurston and Jean Toomer wrote novels. Jacob Lawrence

Monkeys on Trial

"I object to your statement," says Clarence Darrow. "I am examining you on your fool ideas that no intelligent Christian believes."

Jesus said, "Ye shall know the truth, and the truth shall make you free." Think about that quote for a minute.

Now here's another one, this from Thomas Jefferson, who helped separate church and state in the United States when he wrote *Truth is great and will prevail if left to herself.* Separating church and state means that the government can't pick a belief for you, make you go to church, or make you pay taxes to support a certain church. (Governments used to do that; some still do.) Put Jefferson's words in your head and then take yourself to the town of Dayton, Tennessee (population about 1,600). It is 1925.

Mule-drawn wagons and old Model Ts are rolling down the dusty roads into Dayton. Hot-dog and soft-drink vendors seem to be on every street corner. More than 100 reporters have arrived in town. So have photographers and motion-picture makers. A telegraph office—with 22 operators—is set up in a grocery store. A bookseller hawks biology texts. Another sells Bibles. Everywhere there are monkeys: monkey postcards, stuffed toy monkeys, and souvenir buttons that say *Your old man is a monkey.* Dayton has never seen so many people. What's going on?

It's a sensational court case; the best-known trial of the decade. Newspaper headlines are calling it the "monkey trial," and readers and radio listeners all over the country (and in other countries, too) can't seem to get enough of it.

A young schoolteacher is on trial because of what he is teaching in his classroom. Actually, it is modern science that is on trial, and separation of church and state. In Tennessee it is illegal to teach the science of evolution. Evolution traces life on earth through millions of years of development from simple one-celled creatures to increasingly complex plants and animals to humans. (Since apes and monkeys are a stage below us on the evolutionary ladder, jokesters have come up with the monkey label for the trial.)

William Jennings Bryan

Most fundamentalist Christians have a problem with evolution. They believe in the exact words of the Bible, and the Bible says that the world was created in six days and that Adam and Eve—humans—were part of the Creation from the very beginning. This is a disturbing issue, and very serious to many people. Can you be both a Christian and a believer in evolution? Most Christians (but not fundamentalists) say you can. The theory of evolution is accepted as fact at all of our major universities.

But, in 1925, Tennessee fundamentalist Christians have gotten a law passed that says it is *unlawful for any teacher …to teach any theory that denies the story of the divine creation of man as taught in the Bible, and to teach instead that man has descended from a lower order of animals.*

Now that state law is telling citizens what they should believe. The doctrine of a church is being imposed on public schools. It is the opposite of separation of church and state. The First Amendment to the Constitution protects that separation when it says, *Congress shall make no law respecting an establishment of religion, or prohibiting the free exercise thereof.* Because of the new law, Tennessee's citizens are no longer free to study evolution in public school. (Private schools may teach as they wish.)

Many schoolteachers ignore the law and keep teaching what is in their textbooks and what most believe in—evolutionary science. But the American Civil Liberties Union believes the law is unconstitutional. The ACLU, a private organization, was founded in 1920 to protect civil rights in America. The ACLU says it will pay the legal expenses of anyone who wants to test the Tennessee law. In Dayton,

some citizens sitting around in Robinson's drugstore decide to test the law. They ask 24-year-old John Scopes if he would mind being arrested—Scopes teaches evolution in high school. They joke that a trial might put Dayton on the map. It turns out to be no joke.

When William Jennings Bryan learns of the trial, he volunteers to be the prosecutor (in favor of the Tennessee law). Bryan has run for president three times; everyone knows him. He is kind, well-liked, and a fundamentalist. Clarence Darrow volunteers to defend Scopes. Darrow is a brilliant lawyer, a friend of the underdog, and an agnostic (someone who is not sure if there is a God or not).

When the trial begins, it is hot, and the old courthouse is so crowded it acts as if it might collapse; the judge moves everyone outdoors. Darrow is in shirtsleeves and wears lavender suspenders to hold up his pants. Bryan turns his collar inside his shirt, ties a handkerchief around his neck, and cools himself with a palm-leaf fan.

Bryan accuses Darrow of wanting to "slur the Bible."

Darrow says he wants "to prevent bigots and ignoramuses from controlling the educational system of the United States." Darrow puts Bryan on the stand (that isn't usually done to the other lawyer), and asks questions that Bryan admits he hasn't thought much about. When Darrow forces Bryan to say that six days might not be six actual days, Bryan's fundamentalist friends are aghast. The great Populist orator is made to look foolish. (He dies in his sleep five days after the trial ends.)

It is an angry trial, full of bad feelings, and it doesn't settle much of anything. Bryan does win the case: the local court and the state supreme court agree that Scopes broke the law. (Because of a technicality, the case cannot be appealed to the U.S. Supreme Court; the law stays on the books until 1967.) But, in most of the nation, people laugh about monkeys and don't take it seriously. Which is too bad. It is an issue that will keep popping up. In the 1980s, Arkansas and Louisiana pass laws that say that public schools teaching evolution must use "equal time" to teach creationism (the Bible's story of Creation). In 1987, the Supreme Court finds those laws in conflict with the First Amendment's guarantee of religious freedom. Do you understand why? Do you believe that guarantee is important?

Georgia on My Mind

Have you ever known a baker who could take ordinary ingredients, mix them together, and come up with a cake that is unforgettably delicious? Georgia O'Keeffe did something like that with art. She took a single flower, or an old bone, or a mountain, used it for her subject, put a rainbow of colors on her artist's palette, and created paintings that are, in their own way, as luscious as a fresh-baked cake.

O'Keeffe grew up in prairie Wisconsin, went to high school at Chatham Episcopal Institute in traditional Virginia, studied art in Chicago, arrived in New York in 1914 (the year after the Armory Show) for more art studies, married Alfred Stieglitz (one of America's great photographers), and eventually moved to New Mexico (and fell in love with its earth and sky and mountains). "I found I could say things with color and shapes that I couldn't say in any other way—things I had no words for," wrote O'Keeffe. She was right. Check out a book of Georgia O'Keeffe paintings from the library and see what you think of them.

A Writer for the Jazz Age

F. Scott Fitzgerald, a young Minnesotan (named for an ancestor, Francis Scott Key), had a tough time in school—but he loved to write:

When I lived in St. Paul and was about 12, I wrote all through every class in school in the back of my geography book and first-year Latin book and in the margins of themes and mathematics problems. Two years later my family decided that the only way to force me to study was to send me to boarding school. That was a mistake. It took my mind off my writing.

But not for long. In a short life (he died at age 44), Fitzgerald wrote five novels, more than 150 short stories, and many essays. Hardly anyone was able to describe America of the 1920s and 1930s as well as Fitzgerald could. It was Fitzgerald who labeled the times the "Jazz Age." His characters tried to turn life into a glittering party, and usually did, and then couldn't understand why they weren't happy. Something was wrong, Fitzgerald said. His generation had "grown up to find all Gods dead, all wars fought, all faiths in man shaken."

and Romare Bearden painted pictures. And Duke Ellington and a whole lot of other people made music. Artistic excellence was something that the segregationists couldn't suppress. And Harlem, during this time known as the Harlem Renaissance, exploded with creativity.

While this was going on, unsmiling Calvin Coolidge sat in the White House, and people thought him the perfect president. Why? Well, maybe it was because he was dull. "Keep Cool with Coolidge" was his campaign slogan. It was comforting, with all the changes going on, to have a sober, solid citizen as a leader. "The business of America is business," said Cal, and everyone thought that a fine statement.

Coolidge was uncomfortable in front of the camera—in this down-home photo opportunity he looks as if he's never been on a farm in his life.

9 Everyone's Hero

Ruth the rookie found even riding an elevator exciting. "Why, he's just a babe in the woods," his teammates said.

George Ruth, Jr. (right), aged seven, at St. Mary's with his baseball buddy John DeTullio.

George Herman Ruth certainly didn't look like a hero. His body was shaped like a barrel with spindly legs sticking out of its bottom. His face wasn't much to look at either. In the middle was a mashed-in nose.

But none of that mattered, because Ruth—who was known as "the Babe"—turned beautiful when he stepped onto a baseball field. He was the most famous ballplayer of all time.

He had an awful childhood. He wasn't an orphan, as some books say, but his parents, who ran a tavern, didn't care for him much. So he spent most of his time on the streets of Baltimore, got in trouble, and, at age eight, was sent to a Catholic boys' home (where he played a lot of baseball). He was tough—he had to be—but not bitter or angry. Actually, he was funny and friendly, and had all the instincts of a natural ham actor. He loved playing baseball, he broke all its records, and he was always himself, which means no one could predict what he would do or say next. The crowds adored him.

Ruth was a lefty—a southpaw—and he started out as a pitcher (in 1914) with the Boston Red Sox. He was sensational. But he could also hit—harder and farther than anyone. So Boston had him pitch some days and sometimes had him play first base. In 1918 he was about the best left-handed pitcher in the game, and that year he also led the American

In 1911 Ruth was St. Mary's star ballplayer. He could play just about any position—here he is at shortstop.

Black Sox

Babe Ruth was a godsend for a national game deep in scandal. In 1919, members of the Chicago White Sox (forever after known as the "Black Sox") took money from gamblers and lost the World Series on purpose. Only a larger-than-life figure like Ruth could distract the millions of baseball fans from the ugly scandal and restore the game's innocence. Ruth, with his prodigious hitting, flamboyant behavior, and childlike enthusiasm, was just what the country needed.

47

Ruth (second from right) as a Red Sox rookie in 1914. His first professional baseball experience was with the Baltimore Orioles, who were down on their luck and in the minors. The owner was short of cash and sold Ruth to Boston after five months. By 1916 the Babe was the best left-handed pitcher in baseball.

America fell in love with organized sports during the Roaring '20s. Sports stars became American heroes. Working hours were changing, and more Americans had more leisure time. They could go to ballparks or listen to games on radio. (Red Barber was now one of their favorite sportscasters.) They could also play sports themselves. When the Great War ended there were very few tennis courts or golf courses in the nation. Some states didn't have any at all. Those sports were mainly played by rich folks. By the end of the '20s, golf courses and tennis courts were popping up everywhere. Americans were hard at play.

League in home runs.

But Boston needed money, so they traded him to the New York Yankees. The Yankees put him in the outfield. Pitchers play only every three or four days. An outfielder plays every day. The Yankees were counting on Babe Ruth's hitting.

Were they ever right! Before Ruth, baseball had been a low-scoring game. Pitchers were the stars, and batters did a lot of bunting and base-stealing. Ruth made it a hitter's game.

The all-time home-run record—set in 1884—was 27 home runs in one season. In 1920, Ruth hit 54 home runs. People came to the ballpark just to see those homers. Other ballplayers began holding the bat the way he did and swinging with all their might. Scores began going up. High-scoring games are a lot more exciting than pitchers' duels. The fans went wild. Baseball teams agreed to switch to a lively ball that went farther than the ball they had been using. The game got really exciting.

In 1921, Babe Ruth hit 59 home runs and scored a total of 177 runs. Yankee baseball attendance doubled. In 1923, the New York Yankees built big, beautiful Yankee Stadium. They used the money that came from increased attendance. It was called "the house that Ruth built."

Those who couldn't get to Yankee Stadium could still enjoy the game. The first radio station, KDKA in Pittsburgh, began broadcasting in October of 1920. Four years later there were 576 licensed stations and five million radio sets in use. People in Kansas could hear the Yankees play baseball. Radio announcers went to the ball games and described what was happening, play by play. At home, families clustered around their wood-covered radio sets, and, if the static wasn't too bad, listened for the swat of Babe Ruth's bat.

He was trying to beat his own home-run record. But when a reporter asked, "Which one was that, Babe?" the New York slugger was cool. "I'll hit 'em, you count 'em," he said.

In 1927, the Babe hit 60 home runs. That record stood for 34 years, until 1961, when Roger Maris hit 61. (There were 154 games in the 1927 season and 162 games in 1961.) It took 13 more years for someone to top his career home-run total—714 of them. (Atlanta Brave Hank Aaron did that.)

Babe Ruth loved kids. Someone told him that one of his young fans was dying in the hospital. Ruth went to see the boy, gave him an autographed ball, and promised to hit a home run for him that very afternoon. He did it. The boy recovered and Babe Ruth said, "A home run is the best medicine in the world." At least, that was the story the publicity agents told. It was hard to separate the truth about Ruth from the legend. Eleven-year-old Johnny Sylvester was badly hurt when he fell off a horse—but he wasn't dying.

Above, left, the house that Ruth built—Yankee Stadium. Right, the most famous swing in the history of baseball. At Ruth's funeral in 1948, an old newsman said, "I stopped talking about the Babe for the simple reason that I realized that those who had never seen him didn't believe me."

The End of a Streak

The first great baseball game—at least, the first that wowed the press and public—was played in Brooklyn, New York, on June 14, 1870. Some 12,000 fans tried to squeeze into a park built for 5,000 people to see the Cincinnati Red Stockings (the first professional team) play the Brooklyn Atlantics. The Red Stockings had an undefeated 91–game streak; the Atlantics had a lot of grit. They went nose to nose for nine innings until the game ended in a 5–5 tie. How about extra innings? The rules weren't exactly the same as they are today. The game was supposed to be over. But the crowd wouldn't go home—they wanted a winner. So the teams got back on the field, and the pitchers held on, with the tension shifting back and forth, back and forth, until dusk came and the players could barely see. Top of the 12th and Cincinnati got two runs. Things looked desperate for the Atlantics until Joe Start (known as "Old Reliable") came to bat, hit a line drive (with a man on third), and a Brooklyn fan climbed on the back of a Cincinnati fielder (which did slow his throw a bit, just enough to let the runner score). Then right-handed Brooklyn captain Bob Ferguson surprised everyone by batting left-handed (being thus the first recorded switch hitter) and getting a hit. When Cincinnati's first baseman flubbed a grounder, the Red Stockings' winning streak was over.

The Other Babe

When Mildred Didrikson was growing up in Beaumont, Texas, she played baseball with the neighborhood boys. When she began hitting homers the boys couldn't catch, they gave her a nickname. What do you think it was? Why, Babe, of course. When it came to athletics, Babe Didrikson seemed to do it all. She was an all-American basketball player. She was a softball star and pitched in baseball exhibition games against major-league teams. She competed in swimming and diving events and played competitive tennis. She did some boxing. She became a national heroine when she earned three medals in track and field at the 1932 Olympic Games at Los Angeles, California. But she is best known as one of America's outstanding women golfers. Once she won 17 golf tournaments in a row. Babe became Babe Didrikson Zaharias (zuh-HAIR-ee-us) in 1938 when she married wrestler George Zaharias. She has been called the outstanding woman athlete of the first half of the 20th century. No American woman before or since has been a champion in so many sports.

"Is there anything you don't play?" a reporter asked Babe. "Yes," she said. "Dolls."

Ruth was a photographer's dream: he loved to clown for the camera (left, with Yankee mascot Ray Kelly; below, in a movie publicity still). "There was only one Babe Ruth," said a friend.

Still, the public couldn't get enough stories about Ruth, and he did spend a lot of time visiting children in orphanages and hospitals.

Babe did everything in a big way—including eating and drinking. So, when he got sick, with a fever and stomach cramps, reporters said it was because he ate 12 hot dogs and drank eight bottles of soda pop. That wasn't true. He ate like that regularly. But this time he was really sick. It was called "the bellyache heard round the world."

In 1935 Ruth was fat from eating too much. He was 40 years old. Still, there was no stopping him. He was now playing for the Boston Braves, and, in his last professional game, hit three home runs! His final home run was said to be the longest ever hit in Forbes Field.

10 Only the Ball Was White

"Don't look back," said Satchel Paige (left). "Something might be gaining on you." James Bell (right) was pitching for the St. Louis Stars and struck out some tough players. "That kid's cool," said a teammate. So they called him *Cool*. But that didn't seem enough of a name, until someone added *Papa*. It was 1922; he was 19.

Josh Gibson (above) "can do everything. He hits the ball a mile....Throws like a rifle," said pitching whiz Walter Johnson.

Some people said that Josh Gibson once hit a ball over the roof at Yankee Stadium—which was farther than the Babe ever did. As for the unbelievable Satchel Paige, his pitching was so accurate they say he could have stayed in the strike zone pitching to Tom Thumb. Did he have a fast ball? Why, Satchel practically invented the fast ball. Someone who batted against him said that you never saw his pitched balls—just heard the thump in the catcher's mitt and knew they'd gone by. And Cool Papa Bell? Well, Paige himself swore that Bell ran so fast he could turn off the light switch and make it to bed before the light went out.

That lights-out story got repeated as a tall tale and a joke, although it happened to be true. Bell explained that he bet Paige he could do it one night when he learned a light switch was faulty. He won the bet. But usually he didn't need trickery. Cool Papa Bell was fast as Mercury (maybe faster), and Paige and everyone else knew it.

Paige and Bell and Gibson were stars of the Negro Leagues. These were Jim Crow times, when, in much of the United States, schools and ball teams and other things for blacks and whites were separate and unequal. Baseball didn't start that way. The brothers Fleet and

Two Cuban teams, the Cuban Stars and the New York Cubans, added a cosmopolitan touch to the Negro Leagues. Martin Dihigo, who played with the Cubans and also with the Vera Cruz (Mexico) Eagles, ended up in baseball halls of fame in Cuba, Mexico, *and* the United States. In a league where everyone seemed versatile, he still managed to be outstanding. As one player said, "I seen them all for the past fifty years and I still think Dihigo was in a class by himself. He'd pitch one day, play center field the next, and the next day he'd be at first base. Sometimes he even played two or three positions in a single game."

Playing for the Cuban Giants, 1905.

51

Moses (Fleet) Walker played 42 games for the Toledo Blue Stockings in 1884. That was the major-league record for a black player until 1947. A black man, he said, could never become "a full man ...of this Republic."

Welday Walker had played in the old American Association. But when the association died in 1892, Cap Anson (who was a big hitter and a big bigot) made sure that the new leagues fielded whites-only teams. So men of color formed their own leagues.

The Negro Leagues were filled with talented players who played hard and seemed to have a whole lot of fun, too. Not that it was an easy life. Money was usually short, the equipment shabby, the travel brutal, and, in segregated times, blacks almost always had trouble finding hotel rooms or restaurants to eat in. Sometimes they played 200 or 300 games in a six-month season. (Figure the math on that one.) Since they had no ballparks of their own, they had to rent, and the managers wanted to get their money's worth—so, when lights got put in ballfields, they'd often play a doubleheader, and then a night game, too.

What they did best, besides the regular league games, was barnstorm around the country bringing entertainment and fancy ballplaying to blacks (and some whites) in a whole lot of American towns. They were to baseball what the

Cap Anson (back row, second from right) and the Chicago White Stockings in 1888, the year Anson started purging baseball of black players. He seemed to think it was okay to have a black mascot (in front of Anson).

Satchel Paige and his All Stars on a barnstorming tour against white pitcher Bob Feller's All Stars (Paige is in the doorway on the left). "This was a gravy train," said one of the team members. "My share was $7,500 for 17 days' work.... Each team had a private plane. Segregated luxury."

Harlem Globetrotters later became to basketball: wizards. Sometimes they played in formal attire, sometimes in clown costumes, sometimes in uniforms. They did comic routines, they juggled, they did ball tricks. Sometimes they came with a band and a midway. It was carnival, it was fun, and it was skilled ball, too. You had to love baseball to keep up the pace. Josh Gibson and Ted Page remembered playing a twilight game in Pittsburgh, driving to St. Louis for a day game, and then on to scorching Kansas City for a doubleheader. That evening the two of them made it to a hotel porch, where they sat, dog-tired, until they heard some kids playing sandlot ball. Naturally, they couldn't resist that, so they joined the game.

Except for the traveling, most of the ballplayers had a good time, although a few hated having to clown around. They just wanted to play ball. But,

Were the Men Scared?

It was 1931, and the Yankees were playing an exhibition game against the Chattanooga (Tennessee) Lookouts. Chattanooga's owner, Joe Engle (who once traded a ballplayer for a Thanksgiving turkey), had just signed a 17-year-old pitcher, Jackie Mitchell. When the great Babe came to the plate, the new pitcher was called to the mound—and SURPRISE: Mitchell was a *she*.

Her first pitch was low and a ball. The Sultan of Swat swung at the next—and missed. Ditto the next. Babe Ruth didn't like missing anyone's pitch—and a woman's! Babe demanded to see the ball. There was nothing wrong with it. Mitchell wound up, let go, and the Babe watched the pitch fly by. "Stee-rike!" The Bambino threw his bat, stalked off, and the crowd roared. The next batter was big hitter Lou Gehrig. Jackie threw three times. Gehrig swung three times. And that was that.

Some said that Ruth and Gehrig were just being polite to a woman, but no one who was there believed it. Years later, Jackie Mitchell said, "I had a drop pitch and when I was throwing it right, you couldn't touch it." Baseball's commissioner, Judge Kenesaw Mountain Landis, didn't care. He was a misogynist (miss-SODGE-ih-nist—someone who dislikes women). Landis said Mitchell's contract was void, and baseball lost out.

In 1941, Josh Gibson (top, fourth from left), played for Vera Cruz in the Mexican League, hitting 33 home runs. Center, Oscar Charleston, the phenomenal Indiana outfielder, nicknamed the Hoosier Comet. Right, Smokey Joe Williams. Satchel Paige said he was "the greatest pitcher I've ever seen."

53

World Beater

Sometimes Jesse Owens joined the barnstorming Toledo Crawfords. Then the team had a big track star to help draw the crowds. Owens won four gold medals at the 1936 Olympics in Berlin, Germany (no one had ever won four at once before). That displeased the German leader, Adolf Hitler, who believed in Aryan (white northern European) supremacy and thought Negroes were inferior. Owens would race anyone who wanted to run against him (fans got a 10-yard lead). The only person he ever refused to race was Cool Papa Bell, who said he wanted to race around the bases. Owens even raced a horse—and won!

being black, or dark-skinned Cuban, or Mexican, they couldn't join the major leagues. It was crazy. One excuse was that blacks and whites wouldn't play together. But that wasn't so. White all-star teams often barnstormed with the blacks, playing exhibition games. (Black teams won more than they lost.) Everyone knew that players like Oscar Charleston, Smokey Joe Williams, and Buck Leonard were major talents.

"I have played against a Negro all-star team that was so good we didn't think we had even a chance," admitted white Dizzy Dean, who pitched for the St. Louis Cardinals and the Chicago Cubs. "There is no room in baseball for discrimination," said Lou Gehrig (one of the best first basemen ever). White catcher Gabby Hartnett (whose three sisters barnstormed in a women's league) said, "If managers were given permission there'd be a mad rush to sign up Negroes." Which is what finally happened (see Book 10 of *A History of US* for that story). "Democracy is killing Negro baseball," lamented one black sportswriter. But it was a death most people welcomed.

Joe Louis—Making Mom Feel Proud

The Brown Bomber KOs Max Schmeling in one round in 1938.

If [Joe Louis] is not the best boxer that ever lived, he is as near to it as we are ever likely to know. He was born in 1914 on a sharecropper's cotton patch in Alabama and was as country-poor as it is possible to be. In theory the farm was —it had been rented as—a cotton and vegetable farm. But the vegetables did not feed the family, not by the time Joe, the seventh child, came along. His father broke, as sharecroppers do, from the daily strain of not making enough in crops either to feed his children or to put shoes on them. They had no money to send him to a hospital. So he was carried off to a state institution, where he died. A widower came to help out and soon married Joe's mother. And his five children moved in with the eight Louises. Joe got a little more food and went to a one-room school. Then the family moved to Detroit, where the stepfather worked in an automobile factory. Joe went to trade school and worked in the evening doing the rounds with an ice wagon. Then came the Depression, and the family went on relief. This, said Joe, made his mother feel very bad. Years later Joe wrote out a careful check, for two hundred and sixty-nine dollars, which was the amount of relief checks they had had from the government. That, said Joe, made Mrs. Brooks, as she now was, feel better.

—ALISTAIR COOKE, *JOE LOUIS*

11 American Music

Children made music in New Orleans, too, using pans, tin cans, spoons. Louis Armstrong was a street kid who learned to play the cornet at the Waifs' Home.

Do you ever stop and listen to the sounds around you? Have you noticed that indoor sounds are different from outdoor sounds?

Concentrate for a minute. It doesn't matter if it seems noisy or quiet; concentrate hard. You will hear sounds you were unaware of before. What you are hearing is the music of your world. Imagine what the music is like—that natural-sound music—if you are sitting on a California beach near the Pacific Ocean. Then pretend you are skiing in Colorado's Rocky Mountains and listen to the sounds around you. And then open your ears and take an imaginary walk down Broadway in New York City. Think of the differences between city sounds and country sounds, between sea sounds and mountain sounds. Some places, of course, have more sounds than other places.

New Orleans is one of those places with a lot of music in the air. If you look at a map, you'll find New Orleans down in Louisiana at the mouth of the Mississippi River. You can understand that New Orleans has water sounds: sounds from shrimp boats and tankers in the Gulf of Mexico, sounds of water lapping the shores of the city's Lake

There were once about 30 different tribes of Native Americans in the Louisiana region, including the Attakapa, Caddo, Chitimacha, and Tunica.

New Orleans is a hot, sticky town much of the year, and people there have always spent a lot of time out of doors.

The great riverboats that plied up and down the Mississippi were a home for jazz bands—Louis Armstrong played on them—and often the place where people outside New Orleans heard jazz for the first time.

At the Back of the Bus

Sarah and Bessie Delany grew up in segregated North Carolina, and describe what life under Jim Crow was like in their book Having Our Say: The Delany Sisters' First 100 Years *(written with Amy Hill Hearth).*

We encountered Jim Crow laws for the first time on a summer Sunday afternoon. We were about five and seven years old at the time. Mama and Papa used to take us to Pullen Park in Raleigh [N.C.] for picnics, and that particular day, the trolley driver told us to go to the back. We children objected loudly, because we always liked to sit in front where the breeze would blow your hair. But Mama and Papa just gently told us to hush and took us to the back without making a fuss. When we got to Pullen Park, we found changes there, too. The spring where you got water now had a big wooden sign across the middle. On one side, the word "white" was painted, and on the other, the word "colored." Why, what in the world was all this about? We may have been little children, but, honey, we got the message loud and clear. But when nobody was looking, Bessie took the dipper from the white side and drank from it.

Pontchartrain, and dock and riverboat sounds from the great Mississippi.

Because New Orleans has a warm climate, people are out of doors most of the time. So there are people-on-the-street sounds. Today in New Orleans, you can also hear cars, trucks, motorcycles, and airplanes.

Back in 1900, the music of the streets was different. Oh, there were many of the same people sounds—but there were also chickens and pigs in the city, and they scratched and squealed. There was the clippety-clop of horses' hoofs, and the rolling sounds of wooden-wheeled wagons, the hoots of trains, and the sad notes of riverboat foghorns.

Some of the people sounds were different then, too. There were no big supermarkets, so people bought ice, milk, bread, fresh fruit, vegetables, meat, and other things from the backs of the clippety-cloppety wagons. The wagon drivers had to tell people what they were selling. Usually they sang their message. It was the same idea as a TV-commercial jingle, but it might go like this: *I got tomatoes big and fine, I got watermelons red to the rind.* Or like this: *My mule is white, the coal is black; I sell my coal two bits a sack.* Imagine a whole lot of street peddlers all

singing their wares at the same time.

Now, on top of all this, New Orleans had, and has, an unusual mixture of peoples. The city was settled by the French in 1718. (In 1803, President Thomas Jefferson bought New Orleans as part of the Louisiana Purchase.) The French language stuck. Some of the sounds and words of that language can be heard in New Orleans even today.

Many of the French and Spanish men who came to New Orleans in the 18th century married African-American women. Their biracial children, called Creoles, often spoke French or Spanish as well as English.

Out of the sounds of New Orleans, and the mixed heritage of its people, a new music arose. It was American music—unlike anything heard in the world before. It combined the rhythm and drum beat of Africa with the instruments and heritage of Europe. It added a dash from the spirituals of the black Protestant churches, and much from the talents of some black musical geniuses who could be heard in street bands and nightclubs. It was called jazz. It was unique (you-NEEK)—which means totally unlike anything before it. Have you ever mixed red paint with yellow? The color you get is not red, or yellow. It is orange. It is unique. Jazz is like that. It is not African music, or European music. It is uniquely American.

In 1900 nobody much outside of New Orleans had ever heard of it. But in the 1920s jazz began to spread: first to Chicago, then across the country, and then around the whole world.

The best way to learn about jazz is to listen to it. You could start with one of the greatest jazz performers: Louis Armstrong.

Louis was one of those boys who sold coal in New Orleans for two bits a sack. He was very poor. Then someone gave him a trumpet. It must have been a good angel. Louis Armstrong was born to play the

By the mid-1920s, dance halls and speakeasies, like this club scene painted in 1929 by Archibald Motley, Jr., were booming in New York and Chicago. Jazz musicians went north, and soon Chicago had replaced New Orleans as the home of jazz and swing.

The first jazz is said to have been played by funeral bands that wailed music full of soul and sadness as they followed horsedrawn hearses down the streets of New Orleans. It was blues music.

57

Talking Without Drums

Willie Ruff plays the French horn. He knows a lot about jazz. He has taught about it at Yale University. Here is what Mr. Ruff says of the origins of jazz:

In Africa the drum is the most important musical instrument…people use their drums to talk. Please imagine that the drum method of speech is so exquisite that Africans can, without recourse to words, recite proverbs, record history, and send long messages. The drum is to West African society what the book is to literate society.

"In the seventeenth century," Ruff continues, "when West Africans were captured and brought to America as slaves, they brought their drums with them. But the slave owners were afraid of the drum because it was so potent; it could be used to incite the slaves to revolt. So they outlawed the drum. This very shrewd law had a tremendous effect on the development of black people's music. Our ancestors had to develop a variety of drum substitutes. One of them, for example, was tap dancing.…By the time jazz started to develop, all African instruments in America had disappeared. So jazz borrowed the instruments of Western music."

Now you understand that while jazz is not African music, it is mainly the creation of American blacks. You should know, however, that from the beginning, white musicians were playing jazz and some of them made important contributions to the music. While the roots of jazz can be found in Africa, the fruit is a hybrid. (Willie Ruff is wrong when he says no African instruments came to America. The banjo is African in origin.)

Louis Armstrong

Bessie Smith

Many of the early jazz leaders—such as Joe "King" Oliver (back row, on cornet, with his Creole Jazz Band)—were trumpeters, probably because the trumpet could be played while marching, and because its sweet, high sound carried over the street noise. Kneeling, front, is Louis Armstrong.

trumpet. People began calling him *satchelmouth*, because his cheeks seemed to hold a suitcase full of air. "Satchmo" was soon playing on riverboats that went up and down the Mississippi. Then he went to Chicago and began making history.

Satchmo had a big grin, but when he played the trumpet he closed his eyes and blew clear, heavenly tones. Listen to some of his recordings, and see what you think.

As soon as you start listening you will learn something: no two jazz performances are exactly alike. Composers who compose European-style music write down notes and expect musicians to play those notes just as they are written. That isn't so in true jazz. You see,

an important part of jazz is *improvisation*. Improvising means doing your own thing. Jazz musicians talk to each other with their instruments. It is something like African drum talk. One musician leads with a theme. Then someone answers that theme. He plays the theme his own way. Then maybe the first musician improvises with another variation on the theme. Soon the whole band is playing with it. Does that sound wild? It isn't easy to do it well.

People in the 1920s were wild about jazz. The 1920s were called the Jazz Age. When the Jazz Age '20s ended—with a big thud called the Depression—jazz continued to grow in popularity. Today, many people call it America's most original art form.

The Duke Roars with the Twenties

He wore dandy suits, ruffled shirts, top hats, and slicked-back hair. His high-school friends called him "The Duke," because he dressed like royalty. "He was wearing velvet slippers!" remembers one adoring fan. "No man I knew would have dared wear velvet slippers!" The Duke was a man of supreme grace—taking to heart his father's credo that "pretty can only get prettier, but beauty compounds itself."

But beneath the elegance and flash was music. Wild music. Some people called it *jazz*, but not Duke Ellington—that was too small a category. By the late '20s, his music was taking off. It was a time when everyone was desperately trying to be young again. The Great War was over; it had exhausted the world and killed much of its youth. Now that world was ready to party.

Painters came to New York to paint in bold new styles that forsook form altogether. Novelist F. Scott Fitzgerald sketched portraits of the fast life—decadent dreams where characters had such pressing observations as "I like big parties...at small parties there isn't any privacy." Duke Ellington, himself a painter turned composer, created music in colors with songs like "Mood Indigo,"

"Magenta Haze," and "Blue Cellophane."

The sound of the '20s was jazz, originated by blacks but embraced by whites. Innovative jazzmen appeared all over the country. Louis Armstrong showcased his individual talent—many credit him with inventing the jazz solo. Duke's music borrowed from his African heritage, from European atonal theory, and from Western classical tradition—but the music he created was completely his own. It was deeply thoughtful music, sometimes achingly sad, and it reflected the romanticism and gentleness of its creator.

Duke had experienced a comfortable, middle-class childhood in Washington, D.C.—in his own words, he was "pampered and spoiled rotten." He was a man of infinite mystery; as his sister said, "There was just veil after veil after veil." We know that he never liked to argue, that, when confronted with racism, he "took the energy it takes to pout and wrote some blues." His music remains hard to classify; some people still wonder exactly what jazz is. "If you gotta ask," said Louis Armstrong, "you'll never know."

—Danny Hakim
(COURTESY OF *TLC Monthly*)

Duke Ellington

Hooray for Hollywood!

Hollywood, which was the name of a suburb of Los Angeles, was the capital of a make-believe world: nothing was quite real, although everything was made to seem real.

In its golden age, which was the 1930s, Hollywood was a movietown filled with studios where makeup artists, costumers, carpenters, writers, directors, and camera crews created scenes and characters that put dreams and laughter and exotic settings onto a big screen. It was pure enchantment, and it possessed the nation. In later years, the cameras would leave the studios and go "on location" to real places. Then computer technology would make astonishing effects possible; but none of it came close to capturing people the way those Hollywood films did.

Maybe it was the Depression that made people love going to the movies. Maybe cute Shirley Temple, dashing Clark Gable, the comical Marx Brothers, and glamorous Greta Garbo made people forget their troubles. Whatever the reason, some 80 million movie tickets sold each week in 1938—that equaled 65 percent of the U.S. population (although some people went more than once a week, which skewed the figures a bit). Compare it to this: in 1990 about 20 million tickets sold every week, representing less than 10 percent of the population. In the 1930s, movies mattered; now they entertain.

Tyrone Power and Gene Tierney (foreground) in a passionate embrace while filming The Razor's Edge *for Twentieth Century Fox before an intimate audience of at least 50 enthralled grips, continuity girls, lighting designers, cameramen, director, and producers.*

Miss Shirley Temple and Bill "Bojangles" Robinson tap dance down the stairs to fame.

TO HOLLYWOOD IN STYLE

In 1931, it took five days on two trains to get to Hollywood [from New York], but what a deluxe five days! Because that traffic was supported by the film industry, the Twentieth Century *to Chicago and the* Santa Fe Super Chief *[trains] to Los Angeles glittered with polished mahogany, shiny brass, and red brocade; the seats flaunted antimacassars of heavy lace.* —ANITA LOOS, CAST OF THOUSANDS

12 Space's Pioneer

Robert Goddard in his lab in 1935, the year he became the first to shoot a liquid-fuel rocket at supersonic speed.

It was cold in Auburn, Massachusetts, on March 16, 1926. The three men and the woman who stood in an open field wore warm coats, and, as they talked, their breath turned into misty clouds of frost.

That didn't seem to concern them. They were concentrating on a tall structure. It was built of metal pipes. Actually, there was nothing special about the structure. What was attached to it was special. It was a small rocket.

The woman was Esther Kisk Goddard, and she held a motion-picture camera. She would film what was about to occur. Her husband, Robert Hutchings Goddard, a college professor, was in charge. The two other men were his assistants. They were standing in Goddard's Aunt Effie's field.

The rocket had a two-foot-long motor. Pipes ran from the base of the motor to two attached tanks: one tank held gasoline, the other oxygen. Goddard touched a blowtorch to an opening at the top of the rocket. The rocket let out a roar. Slowly it lifted off the launch pad—because that is what the metal-pipe structure was—and rose into the air, speeding upward at 60 miles an hour. Then it crashed into the snowy field. The flight lasted three seconds. The space age had begun.

Robert Goddard and his assistants went back to their laboratory. They were physicists, and New England Yankees, and they understood the value of hard work. They knew they had much yet to do.

They also knew that what they had done that day was as important as what had been done by two brothers from Dayton, Ohio, at Kitty Hawk, North Carolina, in December of 1903. But almost no one else

Esther Goddard's 1929 film of one of Robert's successful liftoffs. Even when the rockets worked, people laughed and called him "Moony" Goddard.

Robert aged about eight with his mother and father, who owned a machine shop, where Robert could examine for himself how things worked.

It took astronomer Percival Lowell 25 years of searching, but in 1930 he found a ninth planet and named it Pluto. It was mathematical investigation of the planet Uranus that made him believe another planet existed in our solar system.

knew that. In 1926, most people would have laughed at a scientist who took the idea of space travel seriously. It would be a long time before the world understood about rockets and rocket travel. Few people ever knew about Robert Goddard.

Maybe that was because he was shy. No, it wasn't just his shyness. It was more than that. He was a loner, and a dedicated scientist. He didn't have much use for publicity. He wanted to get on with his work. He wanted to realize his dream.

He had dreamed of space and interplanetary travel from the time he was a boy. It all began with two books. One was *The War of the Worlds*, which was written by an Englishman, H. G. Wells. The other was *From the Earth to the Moon*, written by a Frenchman, Jules Verne. They are two of the earliest, and best, of the modern science-fiction novels. Goddard read them, reread them, and then read them again.

They made him believe that space travel was possible. In Jules Verne's book, astronauts catapult into space inside a capsule shot from a gigantic cannon. That didn't sound plausible to Robert Goddard. He decided that rockets were the way space would be conquered.

He had three good reasons for that belief.

The launching pad in Aunt Effie's Auburn, Massachusetts, meadow. Goddard's liquid fuels were the same ones that would be used for Germany's V-2 rockets in World War II (more on them later).

Rocket Science

In 1232, at the battle of Kai-Keng, the Chinese had gunpowder-filled tubes attached to long arrows. When the end of the tube was lit, the burning powder made fire, smoke, and a gas that produced a thrust and sent the blazing arrow through the air. Those early rockets were probably not very destructive, but they must have terrified the Mongols who saw them coming.

The Mongols soon produced their own rockets and brought them to Europe, and Europeans began experimenting with them. In England, a monk named Roger Bacon improved them. In France, Jean Froissart launched rockets from inside a tube (like a modern bazooka). Joannes de Fontana of Italy designed a rocket-powered torpedo that zoomed across the surface of the water and set ships on fire. But, by the 16th century, rockets were no longer used in war; aside from their psychological effect, they didn't seem worth the bother.

Automobile age meets space age. At 17, Goddard had a dream about building a space-flight machine. "I imagined how wonderful it would be to make some device which had even the *possibility* of ascending to Mars," he wrote.

Rockets have more push, or "thrust," than other engines of the same weight. Rockets can be designed to hold within themselves all that is needed to make them operate. Since rockets lose weight as they use up their fuel, their highest speeds can come at the end of a flight.

Rockets were nothing new. The Chinese had designed rockets way back in the 11th century. The Chinese rockets were powered by explosive powders. You lit the powder and whammo, the rocket lifted off and the fuel was all used up. Chinese rockets were mostly used for fireworks.

Goddard realized that if a rocket was to soar out of earth's atmosphere and beyond, it would need a fuel that provided steady power, a fuel that wouldn't burn up all at once. A liquid seemed the answer, but liquid fuels demand oxygen in order to burn. There is no oxygen in space. Now, it doesn't sound terribly complicated to figure all that out and build a rocket that carries its own fuel and oxygen tanks. It doesn't seem difficult because we now know it can be done. The first person to come up with an idea isn't sure it can be done. Or how to do it. It took Robert Goddard 10 years of experimenting to build the little rocket that climbed into the sky on that icy-cold March day in 1926.

A few other scientists—in Germany and Russia—were working on the same problem. They worked with theories. Theories are ideas. Goddard was different. He started with theories, but then he tested them. That is the true scientific way.

When he was 38 (six years before the 1926 test flight), Goddard wrote a paper for the Smithsonian Institution in which he theorized that a rocket of 10 tons might be made

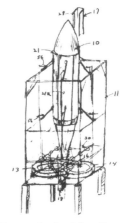

One of Goddard's early sketches for a rocket design.

Sir Isaac Newton (1643–1727), England's great scientific genius, came up with three laws of motion that explain how rockets work and why they can fly in space's vacuum. If you are interested in rocketry, study those laws.

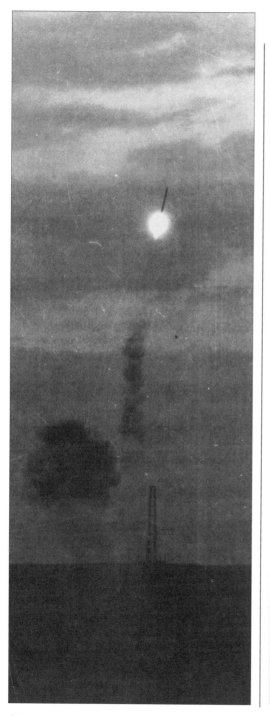

By the 1930s, a Goddard rocket (left) had gone as high as a mile above earth. At right, Goddard (center) with Harry Guggenheim (left), and Charles Lindbergh, who asked Guggenheim's millionaire father, Daniel, to help Goddard; more about Lindbergh on the next page.

powerful enough to reach the moon. He spent much of the rest of his life working on ways to make that possible.

New England isn't a good place to test rockets: too many trees and people. Goddard decided to do his testing in New Mexico, over the desert. Soon his rockets were traveling at 700 miles an hour; some were rising as much as a mile and a half high. He worked out ways to guide and control rockets. He developed the idea of a series of rockets as a means of reaching the moon. He devised parachutes to allow rockets to return to earth smoothly. He patented more than 200 of his ideas.

He didn't live to see his dream fulfilled. Robert Goddard died in 1945, 24 years before Apollo 11 landed on the moon. But his spirit must have been riding with the astronauts on that voyage, probably with Jules Verne and H. G. Wells tucked under its arms.

There is an old Chinese story of a low-ranking but inventive Chinese official named Wan Hu who is said to have designed a rocket-powered flying chair. Two large kites were attached to the chair; attached to the kites were 47 fire-arrow rockets. On the day that Wan Hu decided to test his invention, he sat in the chair and his assistants rushed forward with torches and lit the 47 rockets. According to the story, when the smoke cleared, Wan Hu and the chair were gone—never to be seen again. Today, no one knows what actually happened, but rockets are as likely to explode as to fly.

13 The Lone Eagle

Lindbergh quit college in his second year to go to flying school and buy a World War I biplane.

Robert Goddard might never have gone farther than that field in Auburn, Massachusetts, if a young man hadn't come to visit him. The young man was Charles A. Lindbergh, Jr., and he was famous. Newspapers were saying he was the most famous man on the planet. He may have been.

Lindbergh was wise enough not to be too impressed with himself. He was also wise enough to know the importance of Goddard's work. Lindbergh talked to a philanthropist (fil-AN-thro-pist) named Daniel Guggenheim. (A philanthropist is a person who gives money to good causes.) The Guggenheim Foundation gave Goddard money to build a laboratory in New Mexico.

Why was Lindbergh so famous? Well, back in 1919, a wealthy hotel man offered a prize of $25,000 to anyone who could fly from New York to Paris (or Paris to New York). Several pilots tried for the prize. None made it. Then, in 1927, the competition got fierce. Besides the money, everyone knew there would be much glory for the pilot who first crossed the Atlantic.

In April, Richard E. Byrd took off, crashed, and broke his wrist. (The Byrd family had been well known in America since the days of George Washington and even before.) That same April, two pilots set out from Virginia, crashed, and were killed. In early May, two French aces (top pilots) left Paris, headed out over the Atlantic, and were never heard of again.

In mid-May, three planes were being made ready. Newspapers were full of their stories. The competition had captured the imagination of people on both sides of the Atlantic. Most of the newspaper attention

Byrd Man

Richard Evelyn Byrd was the most renowned American explorer of his generation. In May 1926, he and co-pilot Floyd Bennett flew over the North Pole. (Much later, scientists discovered that his calculations were off and he had missed the North Pole—but not by much.) Byrd was a superb leader and a naval officer who took polar exploration from dog-sled individualism to the discipline of government-sponsored science.

Naturally, one pole led to the other. In 1928, he headed an Antarctic expedition, set up a scientific base there named Little America, and, in 1929, flew over the South Pole. Byrd's radio reports of his adventures were listened to eagerly at home. He wrote books, made lecture tours, and opened Antarctica to scientific research.

65

It was while Lindbergh was an airmail pilot (above, loading the first mailbag ever to fly), flying out of St. Louis, that he met the businessmen who paid for the *Spirit of St. Louis* (inset, top). Right, inset, Paris greets the conquering hero.

In 1922, Bessie Coleman becomes America's first licensed black pilot. Because of prejudice, she has to learn to fly in France.

focused on Byrd, who was famous and eager to try again. His plane had three engines and a well-trained crew. The second plane, with two engines, was to be flown by two experienced pilots. The third plane, a small single-engine craft, could hold only one person. It was called the *Spirit of St. Louis*, because a group of St. Louis businessmen had helped pay for it. The pilot, Charles Lindbergh, was little known. He'd been a barnstormer, a pilot who went around doing trick flying: circles and loops and daredevil things, and then taking people on plane rides for five dollars a spin. That was the kind of thing most pilots did in those days. People didn't use airplanes for transportation. Trains were used to get places. Airplanes? No one was quite sure where the future of aviation lay. But if planes could fly across the ocean safely, they might have an important future.

Lindbergh was a good pilot. He was the first man to fly the U.S. mail

FLIGHT of the SPIRIT of ST. LOUIS

CANADA

ICELAND

ENGLAND
IRELAND

FRANCE
PARIS
⑤ 5:24 N.Y. TIME
10:24 PARIS TIME
SAT.

UNITED
STATES
① NEW YORK
8:00 A.M.
FRIDAY

② 4:00 P.M.

③ 7:00 PM
FRIDAY

3,600 miles

④ 1:30 PM
SATURDAY

Atlantic Ocean

from St. Louis to Chicago. And the first to survive four forced parachute jumps. (Forced because his planes developed troubles.) There was a bold, daring side to him, and another side that was careful and methodical. It was a rare combination. In a crisis he would not panic.

Something about him attracted people. Partly it was his looks. He was tall—six-foot-two—skinny, with light, curly hair and a boyish grin. He looked younger than his 25 years. He was quiet, and was always more at ease with machines, or nature, than with people. He'd grown up in Minnesota, where his father was a congressman. He never did well in school—maybe because he went to a different school almost every year. But he was smart enough to do a lot of reading. He wrote well.

It was eight A.M. on May 20 when he took off. The weather wasn't good, but he was anxious to

I slip Massachusetts into the map pocket, and pull out my Mercator's projection of the North Atlantic. What endless hours I worked over this chart in California, measuring, drawing, rechecking each 100-mile segment of its great-circle route, each theoretical hour of my flight. But only now, as I lay it on my knees, do I realize its full significance. A few lines and figures on a strip of paper, a few ounces of weight, this strip is my key to Europe. With it, I can fly the ocean. With it, that black dot at the other end marked "Paris" will turn into a famous French city with an aerodrome where I can land. But without this chart, all my years of training, all that went into preparing for this flight, no matter how perfectly the engine runs or how long the fuel lasts, all would be as directionless as those columns of smoke in the New England valleys behind me.
—CHARLES LINDBERGH,
THE SPIRIT OF ST. LOUIS

beat the others, and he was used to flying the mail in all kinds of weather.

His little plane carried so much gasoline that some people thought it would never get into the air. But Lindbergh had planned carefully. There wasn't an extra ounce on the plane. He sat in a light wicker chair and carried little besides the fuel, a quart of water, a paper sack full of sandwiches, and a rubber raft. There was no parachute—it would be of no use over the ocean—and there was no radio. He would be on his own once he left the East Coast.

He headed out to sea, and people around the world learned of it on their radios. And then there was nothing to hear. That evening, during a boxing match at Yankee Stadium, the spectators rose and said a prayer for Charles Lindbergh, somewhere over the Atlantic Ocean. Boxing fans are not usually the prayingest people, but that night they were.

Lindbergh, meanwhile, was having a hard time. He had to stay awake or crash. After eight or ten hours of sitting in one place he began to doze. The night before the flight he had been so excited, and had so much to do, that he had not slept at all. So he was tired before he got into the air. He got more tired, much more tired.

Luckily the plane was frail. It banged about in the wind, and each time he started to nod it went careening down toward the water. That woke him. Then, miraculously, the fatigue ended, he looked down, and there was Ireland. He was exactly where the charts—the charts he had drawn—said he should be. Lindbergh, like Columbus, was a superb navigator.

He didn't know that his plane was spotted over Ireland and the news radioed to America and France. People cheered and wept with relief. He was seen over London, and then over the English Channel. Thirty-three and a half hours after he left the United States, he circled the Eiffel Tower in Paris. It had taken less time than he expected, so he was worried that no one would be at the airport to meet him. Since he didn't speak French, he wondered how he would find his way from the airport into Paris. Then he looked down and saw a mob of people. They were waving and screaming.

The young flyer, who had brought nothing with him but the paper bag (which still had some sandwiches), was carried about on shoulders and hugged and kissed and cheered. He was rescued from the crowd only after his helmet was put on another American and the mob thought he was Lindbergh. Then the real Charles Lindbergh was taken to the American ambassador's house, where a butler ran his bath and put him to bed. Soon he was meeting kings and princes and more crowds of admirers. He wanted to stay in Europe and see the sights, but Calvin Coolidge sent a naval cruiser to Europe just to carry him and the *Spirit of St. Louis* back to America. He was now a world hero.

People lost their heads over Charles Lindbergh. All over America there were parades and dinners and celebrations for the man they called the Lone Eagle. It was wild. Nothing quite like it had ever happened before. Why did people go so crazy over Lindbergh?

What he did was daring and brave. But others did daring things. What he did was important. But if he hadn't flown the ocean, someone else would have soon enough. There was more to it than that. The frantic, roaring world of the '20s needed a hero. Lindbergh turned out to be just what was wanted.

You see, he was decent. He didn't drink and he didn't smoke. He was modest. He had good manners. He was offered a great deal of money to pretend to smoke for an advertisement. He wouldn't do it. He wouldn't do anything he didn't believe in.

A publisher offered to hire a ghost writer to help him write a book about his flight. He said he'd write it himself. And he did. He turned down many money-making opportunities. He had values and standards, and he stuck to them.

In the Roaring '20s, when many people thought only of having fun, or making money, or showing off, Lindbergh reminded people that courage, determination, and modesty were perhaps more satisfying.

The American people had found a pretty good hero.

New York City greets Charles Lindbergh in a blizzard of confetti and ticker tape. Later, Lindbergh flew all over the world and pioneered many commercial airline routes.

69

14 The Prosperity Balloon

As secretary of commerce, Hoover tried to restrain big business and the stock market. He failed.

At its peak, in the mid-'20s, the Ku Klux Klan had 4 million members and political clout in the Midwest as well as the South.

Nineteen twenty-eight was an election year. Two good men were running for president. One was Herbert Hoover, an engineer and businessman, who had a reputation for accomplishing things. As a young college graduate, Hoover had managed a gold mine in the Australian desert and then gone to China as a mining expert. After that he got involved (successfully) with Burmese tin and Russian oil. He became known as a very capable man. When Woodrow Wilson needed someone to help feed Europe's starving people (during and after the First World War), he chose Herbert Hoover. He couldn't have found a better man for that job.

Al Smith

The other presidential candidate was the governor of New York. His name was Al Smith, and he was called the "Happy Warrior." Smith was colorful, full of fun, and honest and efficient, too. The governor introduced progressive reforms into New York State, appointed women and minorities to state jobs, and was fiscally responsible. (That means he did a good job with New York's budget.) The Happy Warrior was Catholic, Irish, and a New York City boy.

One of the reasons for reading history is to learn from past mistakes. That campaign was a mistake. Hoover didn't do bad things, but some of his supporters did.

They ran an anti-Catholic hate campaign. It was anti-city and anti-immigrant, too. The miserable Ku Klux Klan was still a powerful force in parts of America. Those are the people who put sheets over their heads and minds. They, and other hate groups, said that the only real Americans were those whose backgrounds were Protestant, English, and white. They said that Catholics, Jews, Asians, Arabs, Germans, Irish, Italians, blacks, and American Indians—of all people—were not real Americans. And then, on top of that, they said they didn't like city people. They said that if Al Smith got elected, the Catholic pope would be ruling America from Rome. Would you believe that nonsense?

Well, sadly, a whole lot of people did.

Hoover was elected by an enormous majority. Perhaps Americans would have chosen him anyway, but the mean-spiritedness of that election should not have happened.

What kind of president was Herbert Hoover? Poor President Hoover. He wore stiff collars and he had a kind of stiff personality, but he didn't deserve the bad luck he had. He worked hard, yet his presidency was a disaster. But anyone elected in 1928 would have been in trouble.

No one understood that then. In 1928 most Americans were rejoicing. America seemed to have acquired King Midas's golden touch. (Although hardly anyone thought to remember what happened to Midas in the end.)

By 1928 the balloon of prosperity had been pumped so full of hot air that no one had ever seen anything quite like it. Many people were saying that something new had been found: an economic balloon that would just keep expanding. A few others were saying: "Stand back, cover your ears, and watch out."

In 1922 this advertisement ran in support of an anti-lynching bill. The bill passed in the House but was defeated in the Senate.

Back to Africa

Marcus Garvey came to the U.S. from Jamaica in 1916 to launch his Universal Negro Improvement Association. Garvey, the most influential black leader of the 1920s, promoted self-help and race pride and urged blacks to return to their African heritage.

The president goes fishin'. "There is no cause for worry," said Hoover's treasury secretary, Andrew Mellon. "The high tide of prosperity will continue."

15 Getting Rich Quickly

The stock market's day of reckoning was October 24, 1929— "Black Thursday."

Oh hush thee, my babe, granny's bought some
  more shares,
Daddy's gone out to play with the bulls and the bears,
Mother's buying on tips and she simply can't lose,
And baby shall have some expensive new shoes.
—SATURDAY EVENING POST

In 1927, '28, and '29 it was easy to get rich. All you had to do was put a little money in the stock market. Then you could watch it turn into a whole lot of money. Here is how it worked.

Imagine that you are the owner of a large company: the ABC Automobile Company. You manufacture cars—good cars—and now you want to expand. You want to add new models. You need to build a new plant and buy a lot of equipment. You need money. So you decide to look for investors. You *go public*. You sell *shares* in your company. You sell 10,000 shares at $100 each. The shares are called *stock*. Anyone who buys that stock becomes a part owner of the ABC Auto Company.

The new automobile models are a big success. The company earns a great deal of money. The stockholders get a percentage of the profits. The money they get is called a *dividend*. The future of your company looks very good; many people want to buy stock in the ABC Company. Here is where a rule of economics, called *the law of supply and demand*, comes in. There are only 10,000 shares available. (That may sound like a lot, but it really isn't.) There is a big demand for ABC stock. People will pay $110 a share for it. Then they will pay $120 a share. Before long, ABC stock is selling for $200.

One stockholder, Mr. Jones, bought 10 shares at $100 each. (How much money did he spend?) Now he will sell them at $200 each. How much money will he have now? What is his profit? (You can do some problem-solving math here!)

Easy to make money that way, isn't it? Well, hold on. In the '20s it was even easier than that. People bought stocks on *margin*. (That means they borrowed most of the money. Today, laws restrict margin buying.) Margin means you don't have to pay $100 for $100 worth of stock. In 1927, you could pay $10, and borrow the other $90 from the stockbroker. (The person who buys and sells stock for you is called a *stockbroker*.)

Now, if Mr. Jones puts $100 into ABC stock and buys on margin at 10 percent, he can have 10 shares instead of one. For $1,000, he will get 100 shares. If he sells the shares for $200 each, how much money will he have? He has to pay back the money he borrowed—$90 per share—but he still makes a whole lot of money. See if you can figure out how much. The good idea that many people soon had was to buy a lot of stock for very little money and get rich quickly.

The business of buying and selling stocks is called the *stock market*. The place where stocks are bought and sold is called a *stock exchange*. The most important stock exchange is in New York City, on Wall Street. Brokers from all over the world call Wall Street with orders to buy and sell.

The stock market usually reflects the business world. If things are

The Securities and Exchange Commission (SEC), an independent government agency, was formed in 1934 to regulate securities (stocks and bonds) markets and investment businesses in the United States. Joseph P. Kennedy was its first chairman. Kennedy had a 17-year-old son, John Fitzgerald, with a big job in his future.

Conducting business at the New York Stock Exchange. A newspaper columnist wrote in 1928, "If buying and selling stocks is wrong, the government should close the Stock Exchange. If not, [it] should mind its own business."

Plans for New York City's Empire State Building (for many years the world's tallest) began at the height of the stock-market boom. By the time it was built, in 1931, half the office floors had to be closed because there were no businesses to occupy them.

Acumen (ACK-yoo-min) is *smarts.*

In 1928 everyone was singing Eddie Cantor's big hit, "Makin' Whoopee." A year later, there wasn't much whoopee around. Depression songs included: "I Got Plenty o' Nuthin'," "Shanty in an Old Shanty Town," the ironic "Life Is Just a Bowl of Cherries," and Eddie Cantor's new hit—"Potatoes are cheaper, Tomatoes are cheaper, Now's the time to fall in love."

going well, stocks go up. If business is poor, stocks go down. Brokers have a nickname for an "up" market. They call it a *bull* market. A down market is a *bear* market. The '20s were a prosperous time. Around 1924, the stock market started rising. At first it was a slow, moderate rise. Then the bulls got frisky.

In 1927, the market began to rise like fury. Almost overnight, stocks doubled and sometimes tripled in value. Everyone was talking about it. Newspapers wrote about it. The country's political leaders were all smiling. They felt it was their good leadership that was causing it. The business leaders were happy too. Of course they thought it was their business acumen that was causing the stock-market boom. Many politicians and business leaders were saying that something new was happening. They said that the boom would just go on and on. That there was not going to be an end to it. And it wasn't just the business and political leaders who were saying all this. Professors from the nation's great universities were saying the same thing.

Now, suppose you are living in 1927. All your friends are getting rich and you aren't. You feel like a dummy, don't you? Why don't you take all your savings and buy as many stocks as you can? Buy them on margin so you can get lots of shares for your money. That's the smart thing to do, say many business experts.

So that is just what you do—in July of 1929. And

Just before he left office, President Coolidge said that the stock market was "absolutely sound." Seven months later, the headlines told a different story.

Times.

THE WEATHER

Rain today and probably tomorrow; somewhat colder tomorrow.

TWO CENTS

STOCK PRICES SLUMP $14,000,000,000 IN NATION-WIDE STAMPEDE TO UNLOAD; BANKERS TO SUPPORT MARKET TODAY

Sixteen Leading Issues Down $2,893,520,108; Tel. & Tel. and Steel Among Heaviest Losers

PREMIER ISSUES HARD

Unexpected Torrent Liquidation Again Rocks Markets.

DAY'S SALES 9,212

Nearly 3,000,000 Shares Traded in Final Hour—1

Inside the Stock Exchange on October 24 was a frenzied, mad scramble as stocks were sold, and then sold and sold again. At 12:30 the Exchange closed its visitors' gallery—it didn't want spectators for this panic. Outside, a huge crowd gathered and rumors of disaster swept Wall Street.

The broker has to sell when the stock price falls unless you come up with more money to pay for the drop in value—immediately.

some of your friends do it, too. The stock boom is phenomenal before July. That summer it is fantastic. Some people are buying stock in anything. It doesn't matter if the company has any real worth or not. No one seems to care. Just give me stock and more stock, I want to get rich. The stock balloon grows bigger and bigger and bigger.

And then guess what happens?

Picture a balloon being pumped up. Now watch that pin. You thought the balloon went up fast? Well, *whoosh*, it will come down much faster.

It happens in October of 1929. It is called the *panic*. People go wild trying to sell. But now almost no one wants to buy.

Remember that $100 stock you bought on margin for $10? Well, when the price drops, your broker sells it. Too bad—you lose your $10. No. It's worse than that.

You see, you owe $90 on that stock. Deduct from the $90 the price your broker gets when he sells the stock. You owe the rest. But you bought 100 shares of stock—or was it 1,000, or 10,000? You put all your

What was the pin that brought down the stock-market balloon in 1929?

After the Wall Street Crash, some people who once had good jobs were now selling apples on street corners. A popular comic strip of the times was called *Apple Mary*. Later, after the Depression, it was renamed *Mary Worth*.

Some Stock Prices

Company	SEPTEMBER 3, 1929	NOVEMBER 13, 1929
Radio Corporation of America (RCA)	505	28
Montgomery Ward	466½	49¼
General Motors	73	36

(Left): a newspaper cartoon for October 25, 1929—"Sold Out." In New York, hotel clerks were said to be asking guests if they wanted the room for jumping or sleeping. Two men, who had a joint account, jumped from a high window at the Ritz Hotel, holding hands.

As fall the leaves by
Autumn blown,
So fell those lovely
shares I own.
Forlorn, disconsolate
I sing,
Goodbye, goodbye to
everything!

To car and plane and
gleaming yacht
And rather ducal
country cot
That all seemed surely
mine by Spring,
Goodbye, goodbye to
everything!
—NEW YORK TIMES,
NOVEMBER 3, 1929

savings into the stock market. The experts said that was the smart thing to do.

Sorry, you owe the money. You don't have any money? You can sell your house, or your car, or both.

You just lost your job, too? You worked at the ABC Auto Company. Oh, that's too bad. Most people have stopped buying cars. Now the company is losing money. It has to sell that beautiful new plant. Why doesn't it sell stock to try to raise money? Are you kidding? No one is buying stock now.

And you know about the banks, don't you? The banks are all in trouble, too. You see, the banks lent money to the brokers and all those people who were buying stocks. Now they have no money. They are closing their doors.

What is happening in America?

We're having a depression, that's what. It will go on for 10 years. People will be out of work. The country will be in terrible shape.

Years later, economic historians will look at the wild stock-market boom and the awful depression and say that they didn't have to happen. If there had been good leadership and sensible regulations, they could have been prevented. That's where greed gets you, they'll say. That's where out-of-date thinking gets you, they'll say.

But where were those economists when they were needed?

And wasn't it fun while it lasted? (For some of us, anyway.) Those roaring, prosperous '20s are great to remember. Didn't we think we were smart that summer of 1929! Will you ever forget what it was like to be rich?

16 Down and Out

The Depression is an embarrassing thing. It is a shame to the system: the American Way that seemed so successful. All of a sudden, things broke down and didn't work. It's a difficult thing to understand today. To imagine this system, all of a sudden—for reasons having to do with paper, money, abstract things—breaking down.

—STUDS TERKEL,
HARD TIMES: AN ORAL HISTORY OF THE GREAT DEPRESSION

There were no jobs for 12 million. Many more had their hours and pay cut.

By 1932, at least 12 million people were out of work. That was one in four of all those who normally would work. Count your friends: pretend that the parents of every fourth person are unemployed. Start with yourself. Suddenly, your family has no income. What are you going to do?

America had had depressions before. They were supposed to be a kind of self-regulating part of capitalism. All the early depressions had something in common: it was the poorest workers who were hurt. They lost their jobs. They went hungry. The wealthy and the middle class suffered only slightly.

The Great Depression (which is what it came to be called) was different. It hurt more people—rich and poor—than any previous depression. And it went on, and on, and on.

To begin, the census of 1920 had shown that for the first time more than half the nation was urban. (Cities in the 1920s were small by today's measures; still, they were a

A depression is a time of decline in business activity accompanied by falling prices and high unemployment. The Great Depression was a time of severe decline in business activity. Today, the government tries to regulate such drastic ups and downs.

U.S. Steel's payroll of full-time workers fell from 225,000 in 1929 to zero on April 1, 1933; even the hands employed part-time in 1933 numbered only half as many as the full-time force of 1929. In Seattle, jobless families whose lights had been cut off spent every evening in darkness, some even without candles to light the blackened room. In December 1932, a New York couple moved to a cave in Central Park, where they lived for the next year.

—WILLIAM E. LEUCHTENBERG

By 1932 about one million people roamed aimlessly across the country, hitch-hiking or riding boxcars. Complete starvation was unusual, but many suffered almost as badly from shame and despair.

big change for people used to farm life.) This was the first major *urban* depression. City people have a terrible time without jobs or income.

Farmers don't have an easy time of it either, but at least they can usually feed themselves. America's farmers, as you remember, had not done well during the '20s. They didn't prosper with the rest of the nation. Crop prices stayed low. So the farmers' income was low too. In 1929, most farms still didn't have electricity or indoor toilets. During the '30s, things got worse. The price of wheat and other grains dropped so low that it was sometimes below what it cost to grow it. Dairy farm-

In 1932, a wagon full of oats did not pay for a pair of shoes. In Illinois in 1933, a bushel of corn sold for 10 cents. There are 56 pounds in a bushel. How much would 25 cents' worth of corn have weighed? Can you see why farmers had problems? Can you see why some were talking about the failure of democracy and capitalism?

ers dumped thousands of gallons of milk onto the land to protest the low price of milk. Other farmers destroyed their own crops. All this waste was happening at a time when city children were hungry. Clearly, something was terribly wrong with our economic system.

Many bankers, brokers,

Some public relief money for the destitute was organized under Hoover—but how did a family of four manage on $5.50 a week?

78

Farmers' income in 1932 was one third of what it had been in 1929. Even farmers who could feed themselves were often unable to pay mortgages, loans, or their taxes. So tens of thousands lost their farms. How would they feed themselves now?

and investors had been wild and irresponsible in the '20s. That irresponsibility caused great hardship in the decade that followed. The American farmer had been irresponsible for generations. Mostly, he hadn't known better, although he should have: European farmers had been practicing crop rotation for over a century.

To *abuse* means to hurt or to treat carelessly.

But, beginning in Jamestown, American farmers had abused land. Farmers used up the fertile land and moved on. They cut down trees and cut up the sod. It didn't have to happen. With careful farming, land can be preserved and enriched. For generations, however, there had seemed to be so much land that few people in America worried. They weren't prepared for nature's tricks: for the droughts and wind

During the 1930s over 3 million people left the Plains to migrate west—the "Okies" and "Arkies" whom the novelist John Steinbeck wrote about in *The Grapes of Wrath*.

The dust storms of the '30s sometimes lasted several days. "It was a terrifying experience," wrote a reporter. "Like driving in a fog, only worse because of the wind, which seemed as if it would blow the car right off the road....a vast, impenetrable black cloud which was hurling us right off the earth....By noon it was as black as midnight."

Evicted means *thrown out.*

storms that came, dried up the land, and turned it to desert. Soil—good, rich topsoil—became dust. Much of the Great Plains just blew away. It was so bad that sailors at sea, 20 miles off the Atlantic coast, swept Oklahoma dust from the decks of their ships. For drought-stricken farmers, there was nothing to do but leave the land, head for a city, and hope to find a job.

But there were no jobs to be found in the cities. City people were moving in with relatives on family farms. It was a time of national calamity.

Now, back to you. Remember, everyone in your family is out of work. How are you going to pay the rent? If you don't have any money, you can't do it. That means you're going to get evicted from your apartment. (If you live in a house and can't pay the mortgage, the bank will take your house.)

What are you going to do? Well, you're lucky: you happen to have a nice aunt and uncle, and they have jobs. So you move in with them. Things are crowded, and everyone gets a bit irritable, but you'll make it.

The family of your best friend—the girl who lives next door—isn't as lucky. They have no relatives to take them in and no place to go. The only thing they can think to do is to build a shack of old boxes and boards on some land near a garbage dump. They are not alone; hundreds of others are camped in that same unhealthy place. Those shanty towns—where people keep warm around open fires—spring up all over the nation. People call them "Hoovervilles," after the president, who says he is trying hard to solve the problem of the homeless and hungry. But nothing he does seems to help. By 1933, a million people in America are living in Hoovervilles.

Top, a famous photographer's glimpse of the fate of the American dream during the Depression. Below, a Hooverville in Seattle, home to thousands of "forgotten" men, women, and children.

Dust Bowl Days

The Dust Bowl is the name given to the region that was devastated by drought during the Depression years. It went from western Arkansas to the Oklahoma and Texas panhandles to New Mexico, Kansas, Colorado, and into Missouri. That area has little rainfall, light soil, and high winds. During World War I (when grain prices were high), farmers had plowed up thousands of acres of natural grassland to plant wheat. When drought struck (from 1934 to 1937), the soil lacked a grassy root system to hold it. Winds picked up the topsoil and turned it into black blizzards. Cattle choked and people fled. The government formed the Soil Conservation Service (in 1935) to teach farmers to terrace the land (to hold rainwater) and to plant trees and grass (to anchor the soil). Artists and writers such as Dorothea Lange, John Steinbeck, and Woody Guthrie photographed, wrote, and sang of the tragedy.

17 Economic Disaster

We have now passed the worst and with continued unity of effort shall rapidly recover. (Herbert Hoover, 1930)
I do not believe that the power and duty of the general government ought to be extended to the relief of individual suffering. (Herbert Hoover, 1930)
We shall soon with the help of God be within sight of the day when poverty will be banished from the nation. (Herbert Hoover, 1932)

When Herbert Hoover is inaugurated, he says, "We in America today are nearer to the final triumph over poverty than ever before in the history of the land."

In many cities, public employees such as teachers had to give up part of their already lean salaries to help pay for soup kitchens.

Mellon pulled the whistle
Hoover rang the bell,
Wall Street gave the signal,
And the country went to hell. —ANONYMOUS DITTY

President Hoover thought the Depression was over. At least that is what he said, and he seems to have believed it. He said the economy was "fundamentally sound." And he said that no one was starving in America.

Well, he was wrong about all those things. All he had to do was look out the windows of the White House and he would have seen hungry people. Thousands of veterans of the First World War were camped in the center of Washington, D.C. They were without jobs. Congress had voted them a bonus for their war service. The bonus was not due to be paid until 1945, but these were difficult times. The men needed the bonus now.

So they marched to Washington to see the president. Some brought their families. They built a Hooverville in Washington, D.C. They had no jobs or money, so they slept in tents, in empty buildings, in shacks on

General MacArthur was helped in his breakup of the Bonus Army by another war-hero-to-be, a Major Eisenhower. Hoover's public image got even worse.

public grass. There were said to be 20,000 of them. They were called the "Bonus Army." Most carried American flags. Some were Medal of Honor winners; some had lost arms or legs in the services. Hoover wouldn't see any of them.

The police asked them to leave. They wouldn't go. So President Hoover sent the army. When the former soldiers first saw the army patrols and tanks and cavalry, they cheered. They thought the troops were parading for them. That was a mistake. The troops came with tear gas, guns, and bayonets. Their leader, General Douglas MacArthur, went farther than the president wished. His troops tore down the shacks; they used tear gas and billy clubs. People were hurt; a baby died. When it was over, Hoover said

What do popular songs tell you about a country? During the Depression people sang a song called "Brother, Can You Spare a Dime?" But the next presidential candidate campaigned with a song called "Happy Days Are Here Again."

The Bonus Marchers' shantytown burns down within sight of the Capitol. Millions watched it on movie newsreels.

The Depression was a terrible spiral. Many people were so poor that they stopped buying goods altogether. If people didn't buy goods, manufacturers couldn't make them. So workers couldn't get work, and they, too, couldn't buy anything. So more and more people became poorer and poorer.

Mrs. Hoover (second from left) gives an old lady a food basket at the Salvation Army to illustrate the virtue of private charity.

he saved the country from mob action. But many Americans hung their heads in shame.

Hoover never understood. Each night he and Mrs. Hoover dressed formally for dinner—in tuxedo and long gown—and were served seven-course dinners by a large staff. Soldiers stood at attention around the table. Hoover thought about cutting down on White House expenditures, he said, but then decided that it would not be good for the people's morale. (The people thought differently.)

Hoover was scared, and so were many other leaders. Democracy and capitalism seemed to be failing. Some people—who were considered to be sound thinkers—were saying that our system was finished. A new day was dawning, they said, and American democracy was out of date. All over the world the disruptions of war and depression were making people turn to dictators. Mussolini had taken control in Italy; Adolf Hitler was gaining power in Germany; and Joseph Stalin was in control in Soviet Russia. Some people in the United States thought those leaders were great men. We now know that they were terrible, bloodthirsty tyrants, but we are historians. We have the advantage of hindsight. We know how things came out. People in 1932 didn't know that.

Charles Lindbergh went to Germany and reported that Hitler was a fine leader. Wisconsin's governor, Philip LaFollette, was an admirer of Mussolini. Many intellectuals were fascinated with communism. They weren't the only ones. Senator Theodore Bilbo of Mississippi said, "I'm getting a little pink [communist] myself." The nation's favorite humorist, Will Rogers, said, "Those rascals in Russia...have got mighty good ideas...just think of everybody in a country going to work."

Could capitalism be saved? Were the democratic ideals of George Washington, Thomas Jefferson, and Abraham Lincoln old-fashioned? Mussolini said, "Democracy is sand driven by the wind." The writer William Manchester said that if Hoover had been reelected, "the United States would have followed seven Latin American countries whose governments had been overthrown by Depression victims." That may be going a bit far. You don't have to agree with every historian. We in the United States have a strong constitution and we value it. Still, this was a revolutionary age. Hoover, who was a brilliant engineer, was not the man for the times. He didn't understand the needs of ordinary people.

In 1930, when many Americans were going to bed hungry, he said: "The lesson should be constantly enforced that though the people support the government, the government should not support the people."

What Hoover meant was that no government money should be spent on relief programs. He thought people could help themselves. If government money was spent it should go to business. That would strengthen the economy, he said, and business money would "trickle down" to the people. Many economists believed as he did.

Hoover didn't realize there was a need for new thinking. People were starving in the land of plenty. The gap between rich and poor had grown big as a chasm.

Of course WE CAN DO IT!

But how could we do it? We couldn't roll up our sleeves and get down to work when there was no work to be had.

Hoover believed in *voluntarism*; he thought individuals should help each other. It was a good idea, and there was a lot of neighborliness during the Depression, but it wasn't enough. More needed to be done, much more, or there might be a revolution.

Henry Ford, who was a pacifist (he didn't believe in fighting wars), bought a gun.

The poet Stephen Spender (who came from England for a visit) wrote about our country:

> Despite its…amazing civilization, the social organization is right back in the Victorian Age.…She has no pensions for her old people; no medical benefits for her workers; no unemployment insurance for any trade.

We Americans had clung to the wrong parts of our economic past. No major industrial nation was so unprepared for calamity.

The United States needed a strong leader, someone who would be open to new ideas. Someone who would take charge. And that was just what we got. Fortunately, that strong leader loved America's democracy. He had been taught that being a good citizen means serving your country.

His name was Franklin Delano Roosevelt, and he became the most loved president since Abraham Lincoln—and also the most hated president since Abraham Lincoln. He was one of the most dynamic, active men in all of American history. And that is amazing, because he could hardly stand up by himself.

When Franklin Delano Roosevelt campaigned for the presidency he traveled all over the country to talk to people.

18 A Boy Who Loved History

James Roosevelt and his son, Franklin. Baby boys wore dresses in the 19th century.

"I think families are the most interesting things in the world," said Franklin Delano Roosevelt. His wife agreed. She said, "In the story of every family is the stuff from which both novels and eventually history is written."

They were right. Every family—rich or poor, famous or little known—has stories. If you don't know your family stories, start asking questions. You are sure to hear some interesting things.

Franklin Roosevelt's family was full of stories, and all his life he listened to them. His father, James Roosevelt, could remember the Civil War. He told Frank stories of Abraham Lincoln and of the general who gave Lincoln a hard time, James's friend General George B. McClellan.

Franklin Delano Roosevelt loved history. He loved the way history can connect you to past times. His own family connections went way back into early American history. The first American Roosevelt was a Dutch farmer named Claes. He arrived in New York about 1650. One of Claes's grandsons started the branch of the family that led to Franklin. Another grandson had descendants who included Theodore Roosevelt and Eleanor Roosevelt (we'll get to her soon).

When Franklin was growing up, his father liked to show him a teapot that had belonged to Franklin's great-great-grandfather, Isaac Roosevelt. Isaac was a banker and a Patriot during the Revolutionary War and a friend of

Franklin with the family dog, Mon▶ did not go to school until he wa◼ Instead he was tutored with the ◼ dren of other local estate owners.

86

Alexander Hamilton. But it was Hamilton's rival, Thomas Jefferson, whom young Roosevelt really admired. He tried to find some connection between Isaac and Jefferson, but he never could. (Years later, when Franklin Roosevelt became president, he tried to be like Jefferson and concern himself with the average citizen, whom he called "the forgotten man." However, he also believed, like Alexander Hamilton, in the importance of a strong federal government.)

Franklin's father wasn't the only one with stories to tell. His mother talked of her family, the Delanos. The first American Delano was French, and a Huguenot, and he arrived at the Plymouth Colony in 1621. (He missed the *Mayflower* by a year.) It was love that brought him to America. He was in love with Priscilla Mullins. She must have been special; Myles Standish and John Alden loved her too. She married Alden and rejected the Frenchman, Philippe de la Noye, who, some years later, married an Englishwoman named Hester. Then he dropped the *ye* from his name and became Delano.

The Roosevelts and Delanos prospered in America. Most of them, like Theodore Roosevelt, became Republicans. But James Roosevelt, Franklin's father, was a Democrat.

When Frank was five years old, in 1887, his father took him to meet President Grover Cleveland. Cleveland was the first Democrat to be president in 28 years. James Roosevelt had contributed money to help get him elected. But being president is no easy job; the day young FDR visited, the chief executive was tired. "My little man," said the huge president to the small boy dressed in a sailor suit standing in front of him, "I am making a strange wish for you. It is that you may never be president of the United States."

Well, you know how kids are. Just tell them what you don't want them to do, and that is what they will go for. So it may have been that day that Franklin Delano Roosevelt first got the idea that he would like to be president.

Franklin Roosevelt, aged four

Hyde Park, where FDR grew up, is a big, comfortable house, but it is homey, not a palace like the neighboring Vanderbilt mansion.

Franklin aged 11, with his mother, Sara. She smothered him with love, advice, and overprotection.

I am interested in and have respect for whatever people believe, even if I cannot understand their beliefs or share their experiences.

—FRANKLIN DELANO ROOSEVELT

87

19 How About This?

Playing sailor, aged six, on the family yacht *Half Moon*, in Maine. Franklin loved ships and sailing all his life.

How would you like this: a house in New York City, a country house overlooking the Hudson River at Hyde Park, New York, and a summer house in Campobello, New Brunswick, by the sea? Now, just for variety you'll take plenty of trips: Paris this year, England the next. You'll be surrounded by servants: cooks, drivers, gardeners, a laundress, and your own private teachers. You'll have a pony and dogs and just about everything you want. In addition, you'll have loving parents who adore you and see that children come to play with you. When you finally go off to school—at age 14—your parents will take you in their own railroad car; it has a bedroom and living room.

Sounds pretty terrific? Well, it wasn't bad—and that was what Franklin Delano Roosevelt's childhood was like. No, he wasn't a prince; he just lived like one. So did other children of the wealthy American upper class in the late 19th century. That was at a time when 11 million of the 12 million families in America had an average income of $380 a year. Only a few thousand families could be called rich. And, compared with the

The Roosevelt house at Campobello in Canada, near the coast of Maine.

really wealthy Vanderbilts or Astors, the Roosevelts were no big deal.

How do you think you would turn out if you had everything you wanted? Do you think you might be vain, arrogant, spoiled, and worthless? Well, that is just how some of those rich kids turned out. (Some poor kids probably turned out that way too.) But not Frank Roosevelt.

Franklin aged three with his dog Budgy and his pet donkey.

His parents gave him good values. They expected him to behave like a gentleman: to be kind, considerate, and honest. They gave him a strong religious faith. (He was an Episcopalian.) That faith gave him courage when he needed it.

He needed courage many times in his life. First, when he went away to Groton. Groton was a rich boys' school (and still is, mostly), but the headmaster didn't believe in pampering. Every boy was expected to take a cold shower each morning. That wasn't really hard, if you gritted your teeth and showered fast; it was much harder trying to be just one of the group when you had always been the center of attention. Franklin was handsome, charming, and friendly—but he never got along with people his age as well as he did with those older and younger. At Groton, he didn't get picked for awards and teams he really wanted, and that hurt; but he had been taught not to complain.

It became a part of his character, not complaining. He would be enthusiastic and act as if everything was fine, even if it wasn't. It made him pleasant to be around, but it also made some of his friends uneasy. They never knew how he really felt.

Franklin went to Harvard, as his parents expected him to do, and to Columbia Law School. His mother intended him to live the comfortable life of a country squire, as his father had before him. After all, with the family money, he had no need to work hard. He didn't have to concern himself with others, but he did. When he was a student himself he wrote to segregated southern colleges appealing to them to do as Harvard did and accept black students.

He may have been concerned with those who were less fortunate than himself because of an important influence in his life, a man he admired more than anyone else. A man who cared about people and wanted to make the world better than he had found it. It was the president of the United States, his cousin Theodore.

When Frank was a child he loved to visit Sagamore Hill, TR's Long Island home in Oyster Bay. There he could romp and run with energetic, fun-loving Teddy and his five children. Frank decided that someday he too would have a big family and play with his family as TR did.

TOP: Franklin aged 17, with his father, who died the next year. RIGHT, in a school play. He loved to have fun and play practical jokes.

Franklin, aged 20, now a big man on campus at Harvard.

One day early that first term [at Groton] a group of older boys... trapped him in a corner of the corridor and ordered him to dance, jabbing hard at his ankles with hockey sticks to make sure he stepped fast enough....Refusing ever to seem a victim, even to himself, he pirouetted and toe-danced in apparent high spirits as if he were part of the fun instead of its object. No one sensed his fear. "He did what he was told," one eyewitness recalled, "with such good grace that the class soon let him go."

—GEOFFREY C. WARD,
BEFORE THE TRUMPET

He also decided he would serve his country—and he set out to do it. First he became a New York state senator. Then Woodrow Wilson made him assistant secretary of the navy. (TR had held that job.) In 1920, he ran for vice president with James Cox against Warren Harding and Calvin Coolidge. He lost, but people began to talk of him as a politician to watch.

He soon had a fine family: a busy wife, a daughter, and four sons. He was an energetic father, full of good spirits, who loved to sail and hike. Yale's football coach watched him exercise and said, "Mr. Roosevelt is a beautifully built man with the long muscles of an athlete."

But what he wanted most of all was to be president. He told that to a classmate while he was still in college. And it looked as if he had a good chance to fulfill his ambition.

Then tragedy struck. He went to bed one night, not feeling well; the next morning, he couldn't move. He was 39 and he had a dreaded disease: *poliomyelitis* (PO-lee-o-my-uh-LY-tiss). Usually it struck children: its common name was infantile paralysis. It was especially hard on adult victims. (Later, Jonas Salk and Albert Sabin developed vaccines to prevent polio; but that was still 20 years in the future.)

Imagine you're an active man, father of five children, with big ideas. Then, overnight, you're crippled. At first Roosevelt couldn't move at all. Slowly, with painful therapy and concentration, he regained the use of his upper body. He would never run again. When he walked it was in heavy braces with agonizing steps. Would you have the courage to go on with your plans? Would you feel sorry for yourself? Would you take it easy and let people wait on you?

At first, everyone was sure his career was finished. Franklin's mother expected him to live the life of a wealthy invalid. But Franklin was determined to live normally. He was a cheerful man, and an optimist. Where others saw problems, he saw challenges.

His parents had trained him not to complain, and, even when he was in great pain, he didn't. As it turned out, he gained something from his terrible illness. It taught him patience and made him more determined. It made him know frustration, and sorrow, and anguish. He—the boy who had had everything—learned to understand those who had troubles.

On top of all that, he was married to a very unusual woman. She, too, was determined that he should not change his goals. Her name was Eleanor.

FDR held his job as assistant navy secretary for seven years. He loved the work, and was disappointed that the navy didn't see much action in the First World War.

20 A Lonely Little Girl

I knew a child once who adored her father. She was an ugly little thing, keenly conscious of her deficiencies, and her father, the only person who really cared for her, was away much of the time; but he never criticized her or blamed her, instead he wrote her letters and stories, telling her how much he dreamed of her growing up and what they would do together in the future, but she must be truthful, loyal, brave, well-educated, or the woman he dreamed of would not be there when the wonderful day came for them to fare forth together. The child was full of fears and because of them lying was easy; she had no intellectual stimulus at that time and yet she made herself as the years went on into a fairly good copy of the picture he had painted.

Even at three Eleanor was lonely. Her mother called her "Granny," because she was "old-fashioned" and not pretty.

Eleanor Roosevelt was writing about herself. She wasn't ugly, but she thought she was, perhaps because her mother was a great beauty. Eleanor had long blond hair, blue eyes, a plain face, and teeth that seemed too big for her mouth. She was shy, easily frightened, and serious. Sometimes she told little lies because she was afraid people would not want to hear the truth.

Her father, Theodore Roosevelt's handsome brother Elliott, was a daredevil horseman and a man-about-town. Everyone who knew Elliott loved him. He was sweet-natured and charming. He became an alcoholic. It ruined his life, and his family's life too.

But he loved his sad-eyed daughter. He called her "little Nell," and told her that he wanted her to grow up to be good, brave, kind, and honest.

Elliott and Anna Hall Roosevelt, Eleanor's father and mother. Her mother was a great society beauty, but cold to her daughter. When she died Eleanor seemed hardly to feel the loss. But when Elliott died, two years later, she refused to believe it. After that he was always alive for her.

Eleanor at her grandmother's. She loved her horse and hated the childish short dresses her grandmother made her wear.

Eleanor's Other Mother

Marie Souvestre was the principal of the school that Eleanor attended in England. She was a strong, proud Frenchwoman, and the girls at Allenswood had to speak French. She taught Eleanor to think for herself, and she taught her that women could become something without the help of men.

And she did. She always remembered those things that were fine in Elliott Roosevelt and forgave him his shortcomings. All her life she kept his letters, read them and reread them, and tried to be the kind of woman who would have made him proud. That wasn't easy.

She had a dreadful childhood. Not an ordinary unhappy childhood; a horrible, awful, terribly lonely one.

She adored her father, but most of the time he wasn't there. She lived for his visits. Sometimes there were wonderful carriage rides with her father driving the horses fast and little Eleanor holding him tightly. Sometimes he promised to see her and then, perhaps because he was drunk, he disappointed his child. Once he took her and his prize terriers for a walk. They stopped at his men's club and he left Eleanor and the dogs with the doorkeeper while he went inside "for just a minute." Hours later he was carried out—drunk; the little girl was sent home in a taxi.

When she was eight her mother died. When she was nine her brother died. When she was ten her father died.

Is this too sad to read? It is all true. Eleanor and her younger brother, Hall, went to live with their grandmother. The grandmother didn't know anything about bringing up children. She lived in a big, spooky house. The noise of children playing bothered her. A governess took care of Eleanor and Hall. The governess didn't like Eleanor. She was mean to her.

Then Eleanor was sent away to school in England. Finally, her life took a happy turn. The principal of the school realized that Eleanor had a good mind and a generous, kind nature. Eleanor became her favorite student. Eleanor soon became everyone's favorite. She had a talent for leadership. She was able to inspire others to do their best.

The child who had not always told the truth, who had struggled against the demons that make people lie, was becoming the kind of person her father had wished her to be. For the rest of her life, Eleanor Roosevelt would be known for her truthfulness.

Eleanor smartly dressed for European travel on a school vacation.

21 First Lady of the World

Eleanor in her wedding dress. FDR's mother made them wait a year to announce the engagement.

Eleanor and Franklin at a friend's house in Scotland on their European honeymoon.

When Eleanor came home from England she was tall and willowy. She still thought she was ugly, but other people didn't. Especially her cousin, Franklin, the most dashing of the young Roosevelts. Like Eleanor's Uncle Theodore, Franklin had energy and ambition. Like Elliott Roosevelt, he had charm. He and Eleanor fell deeply in love.

They were soon married. Before long, Eleanor found herself with five children, three houses that needed managing, and a husband with a busy political career.

Then her husband fell ill with polio. She'd had experience with tragedy. Maybe that was why she handled it so well. Her husband's legs would never carry him again. She said that needn't stop him. There was no reason to change his goals. She could become his legs—and his eyes and ears, too.

They were a team, one of the greatest political teams in history. He became president, but she was his link to the people. He stayed in the White House; she went to coal mines and factories and workers' meetings. Then she told the president what people were thinking.

The first time she spoke before an audience her knees shook. Soon she was one of the most successful speakers of her day. She wrote a newspaper column and a magazine column and books. She was the first First Lady to hold regular press conferences. She served food to needy people, read stories to poor children, visited hospitals, and spoke out for

Eleanor as First Lady, later in life, out and about. Here she is visiting an Ohio coal mine.

The Bonus Army returned to Washington when Roosevelt was president. It was called the Second Bonus Army. He offered the veterans the use of an army camp, sent food, coffee, a convention tent, and doctors, and then added a navy band to entertain them. Mrs. Roosevelt came to visit. "Hoover sent the army," said one Bonus Army man. "Roosevelt sent his wife." Roosevelt did not pay the bonuses, but he did offer the veterans jobs in the Civilian Conservation Corps (CCC), planting trees, creating parks, and so forth. Many went to work.

Marian Anderson sings "America" on the steps of the Lincoln Memorial.

Eleanor's press conferences were generally only for women writers, pushing newspapers to hire more female journalists.

minority rights when few others did.

It is hard for a president—any president—to take time to check out government projects. Besides, everyone knows when a president is coming to visit, so things can be made to look good beforehand. But no one ever knew where Eleanor Roosevelt would pop up. The Secret Service had a code name for her: they called her "Rover." She checked up on government projects and told the president the truth about what was happening. If someone

Eleanor was a fighter against racial discrimination; here, she talks to black educator Mary McLeod Bethune at a conference.

wrote to the president complaining about a problem, Eleanor made sure the letter got answered. Sometimes she invited the letter-writer to dinner at the White House. She invited all kinds of people to White House meals: young and old, rich and poor, people of every race and religion.

Always, she fought for the underdog—for those who were persecuted, or treated unfairly. She wanted to see that all people were given an equal opportunity. She worked for women's rights. She worked for minority rights. She stood on the side of truth and justice.

When a women's organization refused to let Marian Anderson, a renowned black singer, use its auditorium, Mrs. Roosevelt resigned from the organization. She encouraged Marian Anderson to sing on the steps of the Lincoln Memorial in Washington, D.C. Marian Anderson sang "America," and more people heard her than could have fit in any auditorium.

Shy, insecure, ugly-duckling Eleanor had grown up to be a strong, sensitive, capable person. She had a kind of no-nonsense wisdom that made her admired around the world. She has been called the outstanding woman of the 20th century. Eleanor Roosevelt became everything her father wished for—and more.

22 Handicap or Character Builder?

FDR's wheelchair, built from an ordinary kitchen chair.

When Franklin Roosevelt first entered politics he was wealthy, charming, astoundingly good-natured, ambitious, and optimistic—but not very serious. Perhaps everything had come to him too easily. In his boyhood and youth and early manhood he had not known suffering. Now he knew. Now there were steel braces on his legs and a current of steel in his veins.

He had gone through a testing time of great pain when he could hardly move at all. He would never stand by himself again—he always had to call on others for help. Sometimes he crawled to the bathroom. For a proud man it must have been very hard. But if he felt sorry for himself he didn't show it. Even at first, when he was very sick, he made those who came to see him feel good. Everyone remarked on his good spirits, his lighthearted manner, and his great courage.

He was determined to conquer polio, so he worked hard exercising and swimming and learning to manipulate the seven-pound leg braces. His slim, boyish torso became strong, muscular, and powerful. For seven years he stayed away from active political life, trying to learn to walk again,

"How marvelous it feels!" Roosevelt said, the first time he slid into the pool at Warm Springs. "I don't think I'll ever get out!"

95

He was brave, that Roosevelt. O Lordy he was brave. He must have known that he would never be whole, but he was brave. A clear-cut nothing-from-the-waist-down case, and yet he forced himself to walk. With steel and fire, he forced his arms to take him across the room, across the lawn, down the steps. He knew, some part of him knew he would never be walking at the head of the Labor Day Parade again, but he kept on pouring his will into what was left of his muscles, trying to walk that walk again.   —LORENZO MILAM

THE CRIPPLE LIBERATION FRONT MARCHING BAND BLUES

Exuberance is joyful enthusiasm.

Keeping It Simple

Roosevelt knew that good writing is clear and direct. In 1942, during the Second World War, there was fear that we might be attacked by air. It was important that lights not show at night and help enemy pilots spot targets. So a government official wrote this blackout order for Washington, D.C.:

> *Such preparations shall be made as will completely obscure all Federal buildings and non-Federal buildings occupied by the Federal government during an air raid for any period of time from visibility by reason of internal or external illumination.*

Roosevelt rewrote the blackout order. "Tell them," he said, "that in buildings where they have to keep the work going to put something across the windows."

An aide wrote: "We are trying to construct a more inclusive society." Roosevelt changed that to: "We are going to make a country where no one is left out."

Roosevelt had to work tremendously hard to learn to walk with leg braces and crutches. But, said Eleanor, "It gave him strength and courage he had not had before."

but his weak legs would not respond. If he despaired, if he was sorrowful, he didn't let anyone see it.

He refused to act like an invalid. And so he sailed and went to dances and did everything anyone else would have done. Only he did it with more energy and exuberance than most people. At a square dance—where he called the dances—others remarked at what a good time he seemed to be having. This was a man who had always loved to dance. How do you think he really felt?

Perhaps, by not showing his inner feelings, he convinced himself, as he convinced others, that he wasn't hurting.

A man whose office was next door to Roosevelt's mother's house in New York wrote this:

> *Our staff used to watch him from the windows as he got out of his car, clicking the brace on one leg into place, then the other.*

Pulling himself erect by his powerful arms, he would then make his way slowly up the inclined boardwalk which covered one side of the steps. He never failed to pause, grin and wave a greeting to the girls in our windows...as always, every move was a test of courage, met as a matter of course with dignity; he simply would not allow bodily disability to defeat his will.

It would have been easier to become president if he had not had polio. And it probably would have happened. After all, he had the Roosevelt name, that incredible energy, and determination. But his disease changed everything. It made the goal harder to achieve; it made the victory sweeter.

And it made the man different. He had been called a political *dilettante* (DILL-uh-tont), which means an amateur: someone who dabbles, who plays the game for sport, as just one of many interests. There was something to the accusation.

FDR had a Model A Ford built with hand controls. He loved to go out driving; in a car he could feel as physically independent as the next man.

William Phillips, a friend who knew him when he first entered politics, described him as "brilliant, lovable, and somewhat happy-go-lucky...always amusing, always the life of the party." But serious, or focused in his beliefs? No one thought that. "He was not a heavyweight...not particularly steady in his views."

Did his crippling disease make him a stronger, more sensitive, more serious person? There are many who believe it did.

Roosevelt always loved to collect and build model boats and ships, and did so to the end of his life. For a very busy man, he had a lot of hobbies.

23 Candidate Roosevelt

Hunger marchers in Washington, D.C. In 1932, the year FDR was elected president, one out of four Americans belonged to a family in which no one had full-time work.

The little plane tossed about in the heavy wind. The pilot, looking down, followed the path of the old Erie Canal; he was flying from Albany to Chicago. Twice he landed for more fuel. In the back seat, the plane's violent swaying was too much for young John Roosevelt. He threw up.

But John didn't even tell his parents. He knew they were busy. His father was polishing a speech he was soon to deliver. FDR was on his way to the Democratic National Convention to personally accept that party's nomination for president of the United States. No candidate had done that before.

In the old horse-and-buggy days, before telephones, it sometimes took a week or more for a messenger to tell a candidate he had been chosen by his party to run for president. Only then was the candidate expected to make an acceptance speech. That time gap had been continued

---

A New Deal—Deal Us In!

When Roosevelt accepted the Democratic Party's nomination for president, he pledged a "new deal for the American people." A cartoonist picked up the phrase, and *New Deal* was the name soon given to President Roosevelt's domestic (home) policies. The ideas of the New Deal were firmly in the American tradition. They were based on Progressive ideas: on opposition to mo- nopoly; on a belief that government should help regulate the economy; and on the conviction that no one wants to be poor and that most poverty is the result of social problems. The New Deal's methods were experimental; some worked, some didn't. The Progressive Party was important at the end of the 19th and beginning of the 20th century. Theodore Roosevelt was a Progressive.

for tradition's sake. Roosevelt saw no reason to stick with the old ways.

Besides, Roosevelt wanted people to know he would be an active candidate and an active president. If anyone was worried that his weak legs would slow him down, he would show them: they would not. So he flew to Chicago, locked the braces on his legs, and stood before the delegates.

"I pledge you, I pledge myself, to a new deal for the American people," said FDR in his captivating, mellifluous voice. To a nation that had suffered three years of devastating depression, the words *new deal*

Hyde Park's front porch on election night: left to right, daughter Anna, son John, mother Sara, FDR, son Franklin, Jr., and Eleanor.

It wasn't only speculators who lost money in the Great Crash. A lot of banks failed, and many people lost all their savings; there was no banking insurance then.

sounded very good. The Republican candidate, Herbert Hoover, didn't have a chance.

Whether he deserved it or not, Hoover was blamed for the Depression. Roosevelt campaigned hard, but he didn't have to. People wanted a change. The election was a landslide. Forty-two of 48 states went Democratic.

Today, because of the 20th Amendment (adopted in 1933), a candidate elected in November becomes president in January. In 1932, the delay was longer. Roosevelt did not take office until March. Between Election Day in November of 1932 and Inauguration Day in March of 1933, the economic situation in the United States got worse and worse.

By March, the economy seemed close to collapse. Every day more and more banks closed. Gold was being hoarded by those who had it. There was even a question as to whether the government had enough money to meet its payroll. A newspaper reporter described the mood of the people in Washington, D.C., as like "a beleaguered capital in wartime." General Douglas MacArthur prepared his troops for a possible riot. Capitalism, said many experts, was too sick ever to recover.

Mellifluous (muh-LIFF-floo-us) means flowing and sweet as honey. It is from the same roots as *mellow* and *fluid*.

As the black crows of hard times come flying over the horizon, this cartoon mocks Hoover's attempts to scare them off with a straw man.

Beleaguered means under siege.

99

24 President Roosevelt

Over the airwaves, FDR sounded warm, not like a speechmaking politician. He always began a radio talk with the words "My friends."

The patient seems to be dying. Dr. Leave Alone is in the sickroom. Leave Alone is a distinguished man. He wears a dark suit and a dark manner. He has been telling the patient that there is nothing wrong. But the patient is gasping for breath. He screams in pain. The doctor attempts to be calm. "Pull down the shades, keep your voices low, and don't disturb the patient," he tells the family. "Medical science has never encountered an illness like this. We will hope for a miracle." The family weeps, the patient groans. Dr. Leave Alone says, "We have done everything we can," and departs.

Desperate, the family turns to another doctor. A big, hearty man, he comes into the sickroom, pulls up the shades, lets in the sunshine, and in a booming voice tells the patient, "You're going to get well." Then he adds, "The only thing you have to fear is fear itself."

"Medical science may not know how to

"Let me first assert my firm belief that the only thing we have to fear is fear itself," said the new president in his inaugural address. "I shall ask the Congress for…broad executive power to wage a war against the emergency, as great as the power that would be given to me if we were invaded by a foreign foe," he said, and the people cheered.

Dr. New Deal brings an alphabet soup of medicines for sick Uncle Sam. If a remedy didn't work, he would try another.

treat this illness," says confident Dr. New Deal, "but that won't stop us. We'll find a way. First we'll try one method, then another, then another. We'll throw out what doesn't work, we'll keep what does. Start smiling, everybody—you're going to see some action—we'll lick this disease!"

Dr. New Deal, of course, was Franklin D. Roosevelt. The sick patient was the economy of the United States. Almost everyone thought it was dying and would have to be replaced. But not FDR. He was a man of action. Besides that, he was a *pragmatist*. That means someone who believes in whatever will work. Someone who has practical intelligence. That was FDR. He also had energy and enthusiasm, and that gave people confidence. They believed in him. Here are some words to describe FDR:

FORCEFUL
DYNAMIC
RESOURCEFUL
OPTIMISTIC
OUTGOING
SENSITIVE
ENERGETIC
VIVACIOUS
ENTHUSIASTIC
LIKABLE
TIRELESS

When former president Calvin Coolidge was asked for his ideas on how to lick the Depression, he said:

> *In other periods of depression it has always been possible to see some things which were solid and upon which you could base hope, but as I look about me I see nothing to give ground for hope—nothing of man.*

A group of prominent bankers was called to Washington to see what suggestions they had for solving the banking crisis. They had none.

President Herbert Hoover said, "We are at the end of our string. There is nothing more we can do."

That gives you an idea of the gloom and pessimism in Washington on the day Hoover packed his bags and left the White House.

The next day, March 4, 1933, Franklin Roosevelt stood, bareheaded, in front of the Capitol, holding tightly to a

At the Democratic convention in 1932, comedian Will Rogers (right) introduced FDR (far left). He certainly managed to make the candidate laugh uproariously.

Inauguration Day 1933. "America hasn't been as happy in three years as they are today," said Will Rogers (ungrammatically but cheerfully). "They know they got a man in there who is wise to Congress....If he burned down the Capitol, we would cheer and say 'Well, at least we got a fire started anyhow.'"

Thomas Corcoran, a New Deal official, said, "Without bloodshed, the New Deal defanged our most dangerous internal crisis since the crisis of 1861." What did he mean? Do you agree?

101

"THE ONLY THING WE HAVE TO FEAR IS FEAR ITSELF--"

THIS IS ONE RABBIT THAT NEVER FAILED ME!

SPENDING

OLD RELIABLE!

One senator said, "The admirable trait about Roosevelt is that he has the guts to try." His opponents said that programs such as Social Security (right) could be paid for only by raising taxes and spending too much money.

A Woman in the Cabinet

In 1933 President Roosevelt appoints Frances Perkins as his secretary of labor. She serves for 12 years, in all his administrations, and is the first woman cabinet member.

lectern. It was his inauguration day. Some small boys perched in nearby tree limbs; dignitaries sat in special seats; but most of the crowd stood and shivered in the cold wind. When the new president spoke, his strong voice cut through the gloom. All across the land, people clustered around radios to hear what he had to say.

"This nation asks for action, and action now," said President Roosevelt. "We must act quickly."

And that was exactly what he did: act quickly. The first 100 days of his presidency are famous for all the things that got done. Congress was on vacation when Roosevelt took office. He called Congress back into session. He began to act. Soon new programs and laws were pouring out of Washington. "It is common sense to take a method and try it," said Roosevelt. "If it fails, admit it frankly and try another. But above all, try something."

And he also said, "The only thing we have to fear is fear itself."

FDR put together a group of advisers. Newspaper reporters called them "the brain trust." Many were college professors. They were new to government, but they had ideas, intelligence, and a desire to help their country. They worked hard. Washington became an exciting place for idealistic, energetic citizen workers.

Roosevelt's ideas really were a "New Deal." He changed America profoundly. He probably saved American capitalism, but he changed some of its habits. The New Deal did away with most child labor, regulated the stock market, made bank deposits safe, helped make employers pay fair wages to employees, encouraged workers' unions, limited hours of work, helped farmers, brought electricity into rural areas, and gave Americans an old-age pension policy called *Social Security*. The New Deal made the government an active participant in citizens' lives. Yet most of the ideas of the New Deal were not really new. They were the old Progressive ideas in a new package. They had already been tried in Europe. America was behind the times when it came to social welfare.

In order to put people to work, the New Deal sent young people out of doors and paid them to plant trees, build parks, and fight fires. It paid painters to paint murals, writers to write books, and musicians to play and create music. Needy people were given money for food and shelter.

Civilian Conservation Corps members planting seedlings in Oregon. Young men aged between 18 and 25 whose families were on relief got room and board for a year and were paid $30 a month, $25 of which went straight to their families.

New Ideas for a New Deal

Here are some of the best-known New Deal programs (a star means the program still exists today):

• * The SECURITIES AND EXCHANGE COMMISSION (SEC) was formed to regulate the stock market.

• * The FEDERAL DEPOSIT INSURANCE CORPORATION (FDIC) insured bank deposits. We no longer had to fear bank failures.

• The CIVILIAN CONSERVATION CORPS (CCC) gave jobs to more than 2 million out-of-work young men in the nation's parklands: building roads, trails, cabins, and campgrounds. Many are still in use today. "Of all the forest planting, public and private, in the history of the nation, more than half was done by the CCC," says historian William Leuchtenberg.

• The PUBLIC WORKS ADMINISTRATION (PWA) built New York's Triborough Bridge and Lincoln Tunnel; Oregon's Coastal Highway; Texas's port of Brownsville; the road between Key West and mainland Florida; and the University of New Mexico's library.

• The CIVILIAN WORKS ADMINISTRATION (CWA) lasted less than a year, but employed more than 4 million men and women. Opera singers were sent to the Ozark mountains (where none had ever sung before); teachers kept rural schools open; Native Americans restocked the Kodiak Islands with snowshoe rabbits.

• * The TENNESSEE VALLEY AUTHORITY (TVA) began as an experiment in regional planning; it became a corporation to produce and sell electric power and fertilizer.

• * The SOCIAL SECURITY ACT established old-age pensions, unemployment benefits, and welfare benefits for the elderly, children, and the handicapped.

• The WORKS PROGRESS ADMINISTRATION (WPA) was a huge program that put people to work building highways, clearing slums, and doing construction work in rural areas. Writers produced regional guidebooks and did oral histories and other research. Artists decorated hundreds of post offices and other public buildings with paintings and sculptures. Musicians organized orchestras and choruses. Actors toured plays to communities that had never seen live theater before. In four years (1935–1939), the WPA gave jobs to 8.5 million people at a cost of $11 billion.

"Come-along, We're going to the Trans-Lux to hiss Roosevelt."

Before TV, you could watch the president in moviehouses, like New York's Trans Luxe, that showed newsreels: films of current events with voiceover commentary.

New Deal critics saw the government diving gaily into an ocean of spending and debt, taking the drowning taxpayer with it.

Roosevelt did something else. Something that was really new and innovative. Something that people in power almost never do without a battle. He shared power with those who had never held it before.

From the country's earliest days, the leaders of the United States had mostly been drawn from one group: white Protestant men of northern European descent. (That wasn't fair; nor did it reflect the true spirit of the men, like Jefferson, Madison, and Washington, who were responsible for the Declaration of Independence and the Constitution.)

Franklin Roosevelt was part of that white-Protestant-male traditional aristocracy of privilege. But he opened its doors. He included in positions of government power those who had been excluded: women, blacks, eastern Europeans, southern Europeans, American Indians, Catholics, and Jews. He began a process that soon added Muslims, Buddhists, Hindus, and all who are citizens. He rejected the idea of an aristocracy of birth and replaced it with the goal of an aristocracy of talent.

There were those who hated him for doing it. It is hard for us today to imagine how much some people hated him. He was called "a traitor to his class." Some who had gone to school with him refused to speak his name.

Some, in the business world, hated him too. Business leaders had been the heroes of the Roaring Twenties. Calvin Coolidge had said, "The business of America is business." The Depression changed all that. Now Roosevelt was the popular hero, and the American people were demanding that business be regulated for the public good.

Before the New Deal, government had been expected to provide conditions that would help business grow and be profitable. But government was expected to do nothing for the people—the workers—who made business profits possible. When Roosevelt was president, many laws were passed to help workers, farmers and ordinary citizens. Government money was spent on the poor. Some people didn't like that idea. But others understood that, if it was done wisely, the nation would be stronger and better for it.

25 Twentieth-Century Monsters

A magazine portrayed Hitler as the strongman come to deliver the German damsel in distress.

In the Weimar Republic in Germany, the chancellor was chosen (like the prime minister in Great Britain) by his party's leaders—not by direct election. The head of the party with the most seats in the Reichstag became chancellor. Hitler's party was the National Socialist (Nazi) Party.

On the very day of Franklin Delano Roosevelt's first inauguration, the day he told America that "the only thing we have to fear is fear itself," something fearful was happening in Germany. It would change the fate of the world. The Reichstag (RIKES-tahg)—Germany's congress— was deciding to give absolute government power to the German chancellor, Adolf Hitler.

Imagine a country letting its meanest, worst people take charge. Imagine giving those kinds of people the power of life and death over the whole nation. Imagine a nation where children are taught to be tattletales and tell the secret police about anyone who protests—even their parents. Imagine a nation that burns the books of its greatest writers because it fears and hates ideas and truth. Imagine a nation that kills people because it doesn't like their religion or their ideas, or because they are handicapped. That's what happened in Germany in the 1930s.

Germany no longer even attempted to be a democracy. It willingly became a dictatorship—the most evil dictatorship in recorded history. (Although the dictatorship in Soviet Russia was almost as bad.) The Germans used their intelligence and skill to create factories of death. They allowed their government to do unspeakable deeds. Some Germans did not approve, but few spoke out. To do so meant risking their lives.

But in that March of 1933, most people in America paid no attention to what was happening in Germany. The Depression seemed more important. Adolf Hitler? He was a little man with a black brush mustache and dark straight hair that

German industry and transportation collapsed after the Great War. In 1922, these women and children, desperate for fuel, were gleaning coal scraps on a mine dump heap.

105

In October 1922, 4,500 German marks bought one U.S. dollar. In November 1923, the exchange rate was 4.2 *trillion* marks to the dollar. One woman lit the fire with marks; children played blocks with bundles of bills.

fell into his face. He strutted about raising his arm in a straight salute and shrieking his speeches. He didn't seem evil; he seemed silly.

The German people didn't find him silly. They were still angry about the war they had lost. Their leaders and historians had misled them about the causes of the Great War. They had been told that Germany was no more to blame for the start of the First World War than any other nation. That wasn't true. But the Germans believed it; they thought the rest of the world had picked on them. They thought the Versailles Treaty—the treaty that had ended the Great War—was unfair. They were humiliated by the terms of peace. Germany was not allowed to have a large army, navy, or air force. Germany was to make large cash payments—called *reparations* (rep-uh-RAY-shuns)—to the winners to help pay back the costs of the war. Germany was made to say that it was totally to blame for the war.

Since most Germans thought they were no more to blame for the war than others, they were furious, especially about those payments. As it turned out, we lent Germany much more than they ever paid. But most German citizens didn't know that.

Germany's citizens were angry and unhappy. Their country was in awful shape economically. Soon after the war, Germany suffered a time of incredible inflation. The government began printing lots of money (partly to pay those reparations). Printing presses ran day and night. When you print a lot of currency, soon none of it is worth much. Prices in Germany rose beyond belief. In 1923, a Hershey chocolate bar cost 150,000 German marks (in the U.S. the same chocolate bar cost a nickel). German money was almost worthless. Buying a loaf of bread might take a bucketful of bills. People lost all their savings. They had to use up everything they had just to pay the rent. Then, after they had finally got the inflation under control, the Great Depression set in worldwide. Unemployment became a big problem.

What the German people wanted was a leader: someone who could

During the Great War, the German leaders kept telling their people that Germany was winning. So it was a real shock when those same leaders surrendered. Somehow the Germans couldn't believe they had really lost. They felt betrayed.

The Versailles Treaty was harsh; but in 1918, the Germans had forced their own harsh peace treaty on Russia at a place in Poland called Brest-Litovsk.

lead them out of the economic mess, someone who could make them feel good about themselves. During this Depression era people in other countries were looking for strong leaders, too. The Americans chose Franklin Delano Roosevelt. The Germans turned to Adolf Hitler.

They made a big mistake. That mistake would cost them and the rest of the world grief beyond imagining. Their leader was an evil genius who captured his countrymen and women in a web of words and convinced them that he could solve all their problems. He told them that others were to blame for Germany's troubles. He told them that Germany was greater than any other nation and meant to rule the world. He told them that other peoples should be their slaves. He told them that they must love their "fatherland"—Germany—before all else. He told them that they must not worry about right and wrong, because anything Germany did would be right. He told them that *might makes right*—and most believed him.

Hitler wasn't the only one who preached the gospel of nationalism—that loving your nation was more important than loving truth and right actions. Militant nationalism was a disease of the 20th century.

In Japan, a military dictatorship took control of the nation and began stomping on its neighbors. The Japanese, too, were suffering from economic depression. They thought they needed more room for their growing population. In addition, the Japanese were angry at the white-led nations that often acted with arrogance toward

The first Nazi Party annual rally at Nuremberg in 1933. The rallies featured legions of soldiers, blaring music, torchlight parades, and a speech by Hitler as a finale that drove his listeners into a frenzy.

Japanese troops keep captured guerrillas at bayonet point in 1932, during the first Japanese assault on the Asian continent, the invasion of Manchuria.

Summer camp for Hitler Youth, the Nazi children's organization. The pressure to join was very strong.

GERMAN AGGRESSION by 1941.

Francisco Franco

Benito Mussolini

After the Bolshevik revolution in 1917, the old Russian empire became a collection of republics called the Union of Soviet Socialist Republics—the U.S.S.R., or Soviet Union. The republics were not all Russian—they included Central Asian Muslim peoples such as the Uzbeks, the Tadjiks, and the Azerbaijanis; Caucasian peoples such as the Georgians and Armenians; and Baltic peoples such as Latvians and Estonians. These countries were supposed to be independent, but they were not. They were controlled by Russia, the Soviet Union's largest and most powerful republic. Many in America referred to the U.S.S.R. as Russia, or Soviet Russia.

Asian countries. Just as in Germany, the Japanese rulers told the people they were a superior race and destined to rule others. And that was what they began to do. They started by attacking China.

In Spain, a strongman named Francisco Franco muscled his way to power, although many Spaniards (and other Europeans, and some Americans, too) fought against him.

In Italy, a pompous dictator named Benito Mussolini took control of the government. Mussolini was a bully, and, like all bullies, he picked on those

who were weak. He sent Italian forces to Ethiopia. There, Italian tanks, machine guns, and airplanes attacked brave Ethiopians, who fought back with spears and lances.

Joseph Stalin

Russia's dictator, Joseph Stalin, killed millions of his own people—anyone who he believed might threaten his rule. His kind of government, he said, would soon conquer the world.

Mussolini called his political movement *Fascism*. Hitler named his *Nazism*, for National Socialism. In Russia, the forces of evil took charge in the name of *communism*. These were all *totalitarian* forms of government. They were the opposite of democracy. In a totalitarian state, individual people don't matter—only the state is important.

Why did good people listen to these terrible leaders? Why did some modern nations become gangster nations? Those are questions that are hard to answer. It didn't happen here; it didn't happen in England. Were we just lucky, or did our democratic tradition give us the strength to resist the evil thinkers?

The Depression brought grave problems to the people of the United States. In 1933, hogs were selling for only 2½ cents a pound in the Midwest. One farmer had to sell all his hogs to pay his rent for a month. Another farmer sold a wagonload of oats to buy a pair of shoes. Hunger and malnutrition were serious problems in the '30s. Many Americans were angry and desperate. It is not surprising that some of them, too, listened to horrid voices. They needed to blame someone for their problems, so they paid attention to: the Ku Klux Klan; the German-American Bund (BOONT), which was inspired by the Nazis; a radio preacher, Father Coughlin, who spewed out a message of hate; and others. A few Americans no longer believed that "all men are created equal." A few wanted to throw out the Bill of Rights. But most Americans rejected the philosophies of wickedness.

Why did we escape the 20th-century virus of totalitarianism? Was it New Deal leadership? Was it our tradition of liberty and democracy? What do you think? Could it happen here?

Father Coughlin

The Bund combined Hitler worship —its salutes, rallies, and brown shirts—with perverted patriotism: Washington was the "first Fascist."

Charles Lindbergh urged Americans not to fight Hitler. He let his antiwar feelings make him do and say things he would later regret. He allowed himself to be used by the Nazis. The America First Committee that he supported got financial aid from Nazi Germany.

Communism is an economic system, and as such it is not evil. It just doesn't seem to work efficiently. When combined with a harsh political system, as it was in Soviet Russia, it was evil.

109

26 A Final Solution

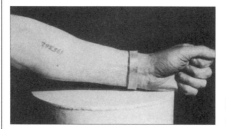

On arrival in a Nazi camp, new inmates' clothes and belongings were taken away, their heads shaved, and identification numbers tattooed on their arms.

In Germany, as soon as the Nazis came to power, Jews were persecuted for no reason except their religion. Jews who weren't even religious were persecuted. A Jewish grandparent was enough to get you into trouble.

There was an evil disease in the world that had been around for a very long time. It was called anti-Semitism. It was hatred of Jews. No one was quite sure where it came from—although the subject was studied endlessly.

No question about it, Jews have been troublesome to some kings and priests. Jews believe that every person is equal before God, which, if you think about it, means that a king is no different from a peasant. Now that is a nonconforming idea. It must have been maddening to some authorities. Imagine where it might lead if everyone thought that way. (Just where might it lead? To democracy, eventually?)

The Jews are people with a powerful book. It is a freedom document, written in Hebrew, and called a *bible*. It tells stories that make people think and ask questions. It tells how, a long time ago, the Jews escaped from Egypt, where they were slaves

Using principles based on "scientific" ideas, Nazis subjected people to absurd tests of their ethnic origins, such as measuring the size of a nose.

of the pharaoh. It tells of Queen Esther and how she saved the Jews when they refused to bow to the king's agent, wicked Haman (HAY-mun). That Bible became the starting point for a new religion called Christianity, and

Berlin, the morning after *Kristallnacht*, November 10, 1938—the "Night of Broken Glass," when the windows of Jewish businesses were shattered by yelling mobs.

for another, called Islam.

At first, Christians were persecuted. Then, in the fourth century C.E., the Roman emperor Constantine became a Christian. Before long, most Europeans were Christians. But not the Jews. The Jews stayed with their beliefs. They wouldn't change their religion even for the emperor or the pope. It was frustrating to those in charge. Others might catch their independent ideas. So some of those in power hated them, and blamed them for whatever was wrong at the time. If there was a plague, it was the Jews' fault; if there was an economic disaster, it was the Jews' fault. Finally, many of Europe's Christians went on religious wars called *Crusades*. The crusaders' aim was to recapture Jerusalem from the Muslims; but anyone not a Christian was considered an *infidel*—a heretic—and was liable to be murdered. Thousands of Jews died, and Jewish property was up for grabs among the crusaders.

In Spain, in 1492, Jews were told they had to leave the country (Jews had already been expelled from many other parts of Europe). They couldn't take their possessions with them. It was a windfall for the rest of the Spaniards. Jews who converted to Christianity were able to stay in Spain, but some were tried by a religious court called the Inquisition, and, if they were found guilty of not being sincere Christians, they were burned alive. That was the opposite of what real Christianity stands for, but most people didn't question the rulers and priests who were in command.

So anti-Semitism stayed in the air. It was still mainly about that nonconformism. Then, in 1517, Martin Luther came into conflict with the Catholic Church and things got complicated. The Catholics, and each of the new Protestant sects, seemed to believe that they alone had the only true religion; that led to centuries of religious wars. Christians were killing Christians—as well as Jews. Hatred and killing in the name of God

---

Freedom of Conscience

In Virginia's Statute for Religious Freedom, Thomas Jefferson wrote:
Be it enacted by the General Assembly, that no man shall...suffer on account of his religious opinions or belief.

John Adams, discussing the subject of religion and government was, as usual, blunt:
Congress shall never meddle with religion other than to say their own prayers.

And James Madison said:
Religion and government will both exist in greater purity, the less they are mixed together.

The Founding Fathers were clear: ours was to be a nation founded on the idea of equality and fairness. There are no religious restrictions on citizenship in the Constitution.

In the Name of Science

That pseudo-science of race led to another "science" called *eugenics*, based on the idea that races should keep their blood pure by getting rid of problems. In the 1880s, Pennsylvania began sterilizing children who were said to be "feeble-minded." (*Sterilizing* means they were made unable to have children themselves.) Indiana was the first state to pass a law forcing the sterilization of certain people who were thought undesirable. (That usually meant the retarded, or criminals, or, often, boys and girls who were sexually active.) Another 31 states followed with sterilization laws. In California, some 17,000 people were sterilized, most of them immigrants. In Virginia, a 1924 sterilization law was challenged, taken to the U.S. Supreme Court—and upheld. In that case, *Buck* v. *Bell*, Carrie Buck had been sterilized because she was said to be retarded. Recent evidence suggests that was not even true. The Nazis used Virginia's law as a model. Under their sterilization act, thousands were made unable to have children.

After 1939, all Jews in Germany had to wear a yellow Star of David sewn onto their clothes.

shouldn't make sense, but it seemed to to some people (who couldn't have been thinking deeply).

There was another factor that produced anti-Semitism. It was economic. Jews were often successful and provided competition. That may have made some people jealous or annoyed.

Then, toward the end of the 18th century, things began to change. After the French Revolution (in 1789), Jews, in one nation after another, were emancipated. They entered Europe's mainstream. People who had been locked in ghettoes were suddenly let out and began a period of great achievement. Especially in Germany, Austria, and Hungary, Jews flocked to the universities, and soon many of them were doctors, lawyers, bankers, store owners, newspaper writers, musicians, teachers, and political leaders. Although Jews made up only about one percent of the German population, they won one quarter of all Germany's Nobel prizes in the first third of the 20th century. Some Germans were proud of that achievement, but others saw it as a problem.

There was something else. It had to do with a science—at least, some people thought it was a science. It was racism, and today we think of it as a false, or pseudo- (SU-doe) science. But, in the 19th century, some thinkers (who believed in what they were doing) divided the world's peoples into races and then said that some races were better than others. They even said that race determines blood, and character, and brain size. They said that the Jews were an evil race that was polluting Aryan (white northern European) blood. They said that people of color were inferior to whites. Since this theory was supposed to be scientific, there were many who believed them.

Hitler used that idea of racism, and bad blood, and the old anti-Semitic virus to explain Germany's problems. It was convenient. Whatever was wrong must be the fault of the Jews. Inflation? Depression? The Treaty of Versailles? It was all because of the Jews, said Hitler. He was an astonishing speaker. People were swept up by his words; they believed him. It was easier than blaming themselves.

The start of the journey for most eastern European Jews, like these people from Cracow in Poland, was a filthy, crowded train journey in a boxcar. The luckier ones ended up in camps where they had to work for the German war effort, such as at this airfield.

Besides, many Jews had good jobs and nice homes. All their property was inviting. Hitler was soon giving it away.

Germany went farther down the road of wickedness than any nation in history. The Nazis used the technology of the modern world for purposes of murder. They built factories for killing (they put all the death camps in Poland). Then they hunted down the Jews of Europe, packed them in railroad cars, and sent them to be slaughtered. They didn't just kill Jews. Hitler hated Slavs (who lived in eastern Europe), gypsies, people who were crippled, and anyone who didn't agree with him. The Nazis killed as many of those people as they could. They enslaved others. It made Hitler and his terror troops feel powerful (and it set an example for other dictators in the future). Because of what was happening to the Jews and Hitler's other victims, all of Europe shivered. People knew that after the Jews were gone it could happen to them.

"The removal and transportation of Europe's Jews was a fact known to every inhabitant of the continent," says John Keegan, a historian of the Second World War. "Their disappearance defined the barbaric ruthlessness of Nazi rule...and warned that what had been done to one people might be done to another."

Guests or Prisoners?

After 1943 (until war's end), no Jews entered the United States except for 874 "guests of the president" who were denied visas, sent to an internment camp in Oswego, New York, kept behind barbed wire, and told they would have to leave the country as soon as the war was over. Some had close relatives in the U.S. One refugee, whose paralyzed wife lived on Long Island, could not visit her even at holiday time.

More than 55,000 immigrant quota spots for eastern Europeans went unfilled in 1944. During the war years, about 100,000 German prisoners of war—mostly Nazi soldiers—were safe in the United States.

Visiting through the barbed wire at Oswego.

113

Arrival in Auschwitz

Auschwitz (OWSH-vits) had been a Polish military barracks, but in 1940 it was turned into a concentration camp. Two years later, gas chambers and furnaces (for killing purposes) were added at a section of the camp called Birkenau. By this time Hitler was frantically rounding up Jews from across Europe; Auschwitz, because it was on a major railroad line between Cracow (in Poland) and Vienna (in Austria), was where most were sent. Day and night, sealed trains arrived from Holland, France, Austria, Czechoslovakia, Yugoslavia, Italy, and other European countries. Auschwitz grew to encompass 40 square miles.

The Wiesels arrived in 1944 on a train from Hungary. One of the family survived: a son named Elie. He wrote a book called Night *that tells what happened to him. Here is some of it.*

The cherished objects we had brought with us thus far were left behind in the train, and with them, at last, our illusions.

Every two yards or so an SS man held his tommy gun trained on us. Hand in hand we followed the crowd.

An SS noncommissioned officer came to meet us, a truncheon in his hand. He gave the order:

"Men to the left! Women to the right!"

Eight words spoken quietly, indifferently, without emotion. Eight short, simple words. Yet that was the moment when I parted from my mother. I had not had time to think, but already I felt the pressure of my father's hand: we were alone. For a part of a second I glimpsed my mother and my sisters moving away to the right. Tzipora held Mother's hand. I saw them disappear into my distance; my mother was stroking my sister's fair hair, as though to protect her, while I walked on with my father and the other men. And I did not know that in that place, at that moment, I was parting from my mother and Tzipora forever. I went on walking. My father held on to my hand.

Children at the camps were photographed in prison uniform so they could be identified if they escaped. Few lived long enough to try.

Behind me, an old man fell to the ground. Near him was an SS man, putting his revolver back in its holster.

My hand shifted to my father's arm. I had one thought—not to lose him. Not to be left alone.

The SS officers gave the order: "Form fives!"

Commotion. At all costs we must keep together.

"Here, kid, how old are you?"

It was one of the prisoners who asked me this. I could not see his face, but his voice was tense and weary.

"I'm not quite fifteen yet."

"No. Eighteen."

"But I'm not," I said. "Fifteen."

"Fool. Listen to what *I* say."

Then he questioned my father, who replied: "Fifty."

The other grew more furious than ever.

"No, not fifty. Forty. Do you understand? Eighteen and forty."

He disappeared into the night shadows. A second man came up, spitting oaths at us.

"What have you come here for, you sons of bitches? What are you doing here, eh?"

Someone dared to answer him. "What do you think? Do you suppose we've come here for our pleasure? Do you think we asked to come?"

A little more, and the man would have killed him.

"You shut your trap, you filthy swine, or I'll squash you right now! You'd have done better to have hanged yourselves where you were than to come here. Didn't you know what was in store for you at Auschwitz? Haven't you heard about it? In 1944?"

No, we had not heard. No one had told us.

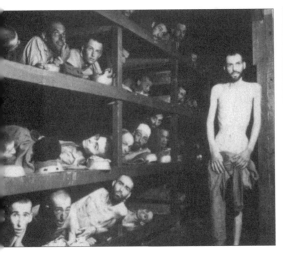

Did all this have anything to do with the United States?

That is a good question.

Suppose you see someone beating up someone else. Really beating her up. She is going to end up in the hospital, or maybe dead.

What do you do? This isn't your fight. If you try to break it up you almost certainly will get slugged. Should you call the police? Should you stay out of it? If someone is killed it won't be your fault. Or will it?

Are we responsible for others? Do you agree that "He who permits evil, commits evil"?

In 1939, 20,000 children—all under 14—were in danger in Germany. Hitler wished to get rid of them. He was willing to let them leave the country. Many were Jewish; some were not. Quakers, Jews, Catholics, and members of other American church groups agreed to take responsibility for them. It would not cost the government any money. Surely America would accept them.

Top left: laborers at the camp at Buchenwald. Many starved to death or were killed when too weak to work. Top right: inmates in an Austrian camp where the Nazis were said to perform "scientific" experiments on human beings. Center: a gas chamber at Maidanek, in Poland. New arrivals were told to undress for showers; poison gas came out of the shower heads. Left, a box full of the wedding rings that all were made to remove.

This is the land of promise. The land built on a spirit of generosity. The land that, from the days of the Pilgrims, has been a place of refuge for the persecuted of other nations.

115

The League of Nations had given Jews the right to buy land in Palestine, but the British put pressure on European nations not to let them emigrate to Palestine. In 1942, the ship *Struma*, with 769 refugees aboard, was turned away from Palestine by the British. The ship sank in the Bosphorus; one passenger survived.

Many Jews and opponents of the Nazis did escape, some to the United States. Among them were pianist Rudolf Serkin (top) and philosopher and writer Hannah Arendt (bottom), who wrote an important book about Hitler and fascism called *The Banality of Evil.*

Those who had founded this nation, and written its constitution, had been clear about it. Although they were much alike—white, male, Protestant, and of English descent—the Founders didn't limit the nation to people like themselves. For they were unselfish in spirit and very wise. They believed they were creating something new on earth, a generous nation that would find strength in diversity. A nation that would take peoples from all over the world and allow them to become a new people—an American people—more varied in its roots than any before it.

The Founders offered the gift of citizenship, not just to their kinsmen, not just to the strong, or the handsome, or the rich, but fully and equally to all who came here to live.

The nation grew, and its citizens understood what was intended. Under the Statue of Liberty they carved the words *Give me your tired, your poor, your huddled masses yearning to breathe free.* Men, women, and children—from all over the world—came to this "promised land." Many were *refugees*—people fleeing tyranny and persecution.

Of course the children Hitler was threatening with death would be welcome in America. Or would they?

There was another tradition here. It was not the tradition of Jefferson, Washington, and Madison. It was a spirit of greed and selfishness.

It was that selfish spirit that had caused Alien and Sedition acts to be passed soon after the nation was founded. It was that spirit that had caused men to rise in Congress and say that slavery was a "positive good." It was the spirit behind the Know-Nothing Party and the nativists. It was the spirit of the Ku Klux Klan and the only-one-race-allowed country club. It was mean-spirited. It was anti-American, but it was there.

Here is what Abraham Lincoln wrote in 1855:

> As a nation we began by declaring that "all men are created equal." We now practically read it, "all men are created equal except Negroes." When the Know-Nothings get control, it will read "all men are created equal except Negroes and foreigners and Catholics." When it comes to this, I shall prefer emigrating to some country where they make no pretense of loving liberty.

The Imperial Wizard of the Ku Klux Klan, in 1923, said: *Negroes, Catholics, and Jews are the undesirable elements in America.* The Imperial Wizard was a bigot, but some people listened to him. They didn't know their history. Maybe they hadn't read the famous letter George Washington wrote the Jews of Newport, Rhode Island. In it, he said: *The government of the United States…gives to bigotry no sanction, to persecution no assistance.*

In 1924, Congress passed a racist immigration bill. Its aim was to keep Asians, Jews, blacks, and people who can't speak English out of America.

Anti-Semitism and xenophobia (zen-uh-FO-bee-ya—anti-foreignism) had infected some Americans. Some of the people with the disease were in Congress, the State Department, and other government offices. Did they realize they were being un-American?

Most Americans are hospitable. Eighty-five newspapers wrote editorials urging Congress to pass a bill letting in those 20,000 children from Germany. Citizens offered their homes to the young refugees. Leaders of church, labor, and social organizations spoke out. But not loudly enough.

The head of a powerful group, the American Coalition of Patriotic Societies, told Congress to "protect the youth of America from this foreign invasion." He shouted the message of the racists. There was fear in the world, and a depression, and Congress listened.

Does this have anything to do with you? Isn't anti-Semitism a Jewish problem? No. It is a human problem. People who hate become hateful. A nation that allows bigotry and persecution is diminished by it. In 1939, the United States government gave sanction to bigotry and assistance to persecution. Those children were not allowed into the United States.

Hitler now knew that no one would rescue the children. He felt free to build death camps. That is where most of those 20,000 children—and a girl named Anne Frank—ended their lives.

The *St. Louis* Is Turned Back

Liane Reif-Lehrer, pictured here (left) in 1938 with her mother and brother, was born in Vienna, Austria, in 1934. These are her words:

I should have been a normal little girl, happy with my special doll, my big brother (who thought me a noisy nuisance but loved me anyway), and my doting parents. But the world around me was going mad, and the life I should have had was not to be.

Liane's father was a dentist, but the Nazis would not let him or other Jews work. One day he was found dead at the bottom of a stairwell. Did he commit suicide or was he murdered? Liane has never been quite sure.

She was not yet five when she, her mother, her brother, and 934 other Jewish passengers set sail for Cuba from the port of Hamburg, Germany, on the luxury liner *St. Louis*. The passengers all had Cuban entry permits, and most had quota numbers that would have let them into the United States eventually. But while they were at sea, Cuba changed its immigration policy. Most of the passengers were not allowed off the ship.

After days of frantic negotiations, the *St. Louis* was forced to leave Cuba. The captain didn't want to take the Jewish passengers back to Germany; he knew what would happen to them there. He headed for Miami. Telegrams were sent to President Roosevelt asking him to grant asylum to the refugees. The telegrams were not answered. A U.S. Coast Guard boat stayed close to the *St. Louis* to make sure no passengers jumped overboard and tried to swim ashore. Finally, the captain could do nothing; he headed back to Europe. Here are Liane's words again:

My mother and brother and I were among the passengers who survived—about a fourth of those on the ship. We were sent back to Europe and given haven in France, only to find the Nazis at our doorstep again a few months later. But somehow we managed to get to the U.S. in 1941. I was seven, a wide-eyed, bewildered girl, greeted by New York children playing street games to wartime hate ditties. I remember a particularly popular one: "Whistle while you work, Hitler is a jerk, Mussolini is a meanie, and the Japs are worse."

I tried to explain that some Germans were good and some bad. On more than one occasion my "non-groupy" response earned me the wrath of the parents, who did not hesitate to suggest that I "go back to where I came from." It hurts even now when I remember the tone with which those words were delivered.

Dr. Liane Reif-Lehrer became a research scientist and then a consultant. She lives in Massachusetts and is writing a book that tells her story.

27 War and the Scientists

Einstein (left) with American scientist Charles Steinmetz, on a visit to the U.S. in 1921, shortly before he won a Nobel prize.

Albert Einstein was a mathematical physicist who ranks with Galileo and Newton as one of the great thinkers who have helped us understand the universe. He published his theories of relativity in 1905 and 1916. In 1921 he won a Nobel prize. In 1933 he wrote *Why War?* with Sigmund Freud (the father of psychology, and also a Jew).

There were some people who felt they had to see Roosevelt. But the president couldn't see everyone who wanted to see him—especially with a war on.

It was a group of scientists. They had something very important to tell the president. How could they get to him? If they had gone to Mrs. Roosevelt they might have had no problem. But they didn't think of that. They met and planned and worried.

Then someone got a bright idea. There was one scientist whom Roosevelt would listen to. Almost anyone in the world would listen to him. He was Albert Einstein, and he was the greatest scientist of the 20th century and one of the greatest scientists of all time. He had discovered the theory of relativity. That theory changed the way science looks at the world.

Einstein was born in Germany. Because he was Jewish he had to escape from that country, and he did. He became a fellow of the Institute for Advanced Study in Princeton, New Jersey, and an American citizen. Other scientists fled the evil regimes in Germany and Italy, too. Hitler lost their fine minds and talents. Britain and the United States gained them.

Some of those scientists knew that the Germans were working on a secret weapon. It was more powerful than anything the world had ever known. If the Germans developed it, they would probably win the war and rule the world. Roosevelt had to be told.

Einstein arrives in America in 1933, one of the earliest refugees from the Nazi regime.

Italian physicist Enrico Fermi (right) meets Maria Martinez, the great potter of San Ildefonso in New Mexico, near Los Alamos, where the secret research was carried out.

But back then no one paid much attention to scientists who talked of secret weapons. They sounded like dreamers. Roosevelt was busy with important practical matters. He needed to strengthen the army and navy. He needed warships and carriers and tanks and planes. He had a depression to fight, too. Secret weapons? Super bombs? Wasn't that the stuff of science fiction?

The scientists got Einstein to write a letter to the president. Then they got Alexander Sachs, a businessman and an economist (he had advised the government about the New Deal), to deliver it to him. Sachs told the president about the German's secret weapon. The president didn't seem to be paying attention. He must have had all those other things on his mind. Sachs was desperate. How could he get the president to pay attention?

Finally, he remembered that Roosevelt loved history. He used that knowledge. He told Roosevelt that when the French emperor Napoleon was fighting the British, the American scientist Robert Fulton had gone to Napoleon with his steamship invention. Fulton told Napoleon that troops could be carried in steamships. Napoleon didn't pay attention. Roosevelt knew that if Napoleon had listened he might have been able to invade England and win his war.

Roosevelt didn't want to make the mistake Napoleon had made. He was ready to listen to the scientists. He was ready to read Einstein's letter. He decided that the United States should work on the secret weapon and try to develop it before the Germans did.

It was an enormous decision. It would be very costly. It was a race against time. The scientists told Roosevelt that they intended to split tiny particles of matter—atoms—and that would release vast amounts of energy. They could use that energy to make the most powerful weapon the world had ever seen. Roosevelt was convinced. The president made the decision to let the scientists go ahead with their plans. The project was top secret. Not even the vice president knew about it.

Left: Hungarian John von Neumann helped create the first big computers; he also worked on the Manhattan project, as the secret weapon research was called. Right: Hans Bethe, from Germany, who had the ability to find simple ways to deal with complex problems.

Center, the Hungarian-born physicist Edward Teller, one of the central figures in the history of atomic fission.

Einstein regretted forever his part in the atom bomb's development. "I made one great mistake in my life," he said, "when I signed the letter to President Roosevelt recommending that atom bombs be made…but there was some justification—the danger that the Germans would make them." Why do you think he felt like that? (Chapter 42 has some clues.)

1942: Enrico Fermi, Eugene Wigner, and other top scientists get together in a squash court under the stands of a University of Chicago football field to create the world's first controlled nuclear reaction from 50 tons of uranium and 500 tons of carbon. No football game has produced that kind of power!

Spectators at the squash-court nuclear reaction toasted its success with this bottle of wine—out of paper cups.

119

28 Fighting Wolves

In an illustration for an anti-Axis propaganda film, *The Fruits of Aggression*, Hitler, Mussolini, and Tojo lick their lips and slice up a juicy watermelon world.

You are probably wondering how three nations—Germany, Italy, and Japan—could be a threat to the whole world. Have you ever thought about how a wolf terrorizes a flock of sheep? A lone wolf doesn't attack a big flock. He picks them off one by one. Give him enough time and he can kill them all.

Germany, Italy, and Japan were wolves. They were powerful. They thought they could devour the world's nations, one by one. They believed most other countries—especially the democracies—were weaklings.

They had good reason to believe that. In a democracy, everyone's ideas are heard. Sometimes democracies have a hard time acting quickly, because so many individuals and groups are debating each other. In the 1930s there were strong *isolationist* voices in America. They said the oceans—Pacific and Atlantic—protected us from danger. They said we didn't need to pay attention to what was going on in the rest of the world. Some of the isolationists were selfish. They didn't

Isolate means to separate from others. Isolationists believe a nation should stay out of world affairs. Some of the World War II isolationists were the same people who prevented the U.S. from joining the League of Nations after the Great War.

War in Europe suddenly made the Atlantic seem narrower. When the Germans invaded Poland, FDR said, "It has come at last. God help us all."

American mothers protest Lend-Lease, a program to lend equipment and raw materials to help the Allies fight the war in Europe. The mothers feared a war that might include their sons. But there was no escape from this war.

even want to help victims of the war.

Others, who were *pacifists*, didn't think it right to fight any war. They believed that if we behaved peacefully other people might do the same.

Still others—in the military—were attached to old ways of thinking. They thought that battleships could protect us. Our battleships were huge. Some were 800 feet long. Imagine three football fields. (A football field is 300 feet long, so chop off a bit.) Float that picture, and add a crew of about 2,800 men, and guns that fire shells 20 miles or more, which was farther than any other weapon of the time. Battleships were much feared.

A few voices disagreed. They said that air power had changed all the rules of war. The oceans were no longer enough protection. Colonel William ("Billy") Mitchell of the U.S. Army said we needed to build up our air force. He said we needed to build aircraft carriers for our navy. An aircraft carrier is a floating airfield that carries its own airplanes. It is really big. (A bit longer than a battleship, and much wider.)

Mitchell pestered everyone: congressmen, army officers, naval officers, newspapermen. They got annoyed. Only a few people thought that air power was important. Because he criticized his superiors in public, Mitchell was finally court-martialed and thrown out of the army. Some people said his ideas were laughable.

The official program of the Army–Navy football game, in November of 1941, showed a picture of the battleship *Arizona* with this caption: *Despite the claims of air enthusiasts no battleship has yet been sunk by bombs.* (That was meant as a slap at Billy Mitchell and those who agreed with him.)

Top, Billy Mitchell as General Pershing's chief of air services in the Great War. He tried to prove the effects of air power by bombing obsolete warships (right, the *Alabama* in 1921), but no one was interested. They said it would cost too much.

Perhaps not even Germans were born to goose-step—a new recruit gets a lesson.

In 1942, a young architect named Albert Speer took control of German industry. Arms and munitions production, until then feeble and inefficient, increased hugely.

The United States had become weak militarily. We listened to the isolationists. It was partly for a good reason: we hated war. In 1941, our military force ranked 19th in the world, smaller than that of Belgium. At the same time, the armies and navies of Germany, Italy, and Japan had become strong.

Once its economy had recovered from the terrible effects of the Great War, Germany ignored the Versailles Treaty. It built a powerful army and air force. It turned out hundreds of submarines. Japan's naval fleet was awesome. Only a few people seemed concerned. "War could have been prevented," said a British statesman named Winston Churchill. "The malice of the wicked was reinforced by the weakness of the virtuous."

Roosevelt understood that the totalitarian powers were dangerous. He knew they hoped to rule the world. He took them seriously. The president wanted to build up our armed forces. It wasn't easy to fight the isolationists in Congress. He began by sending war supplies to England. That got our factories going. But we were still behind most other nations, and way behind Germany, Italy, and Japan, who were making plans to divide the world among themselves, and were known collectively as the *Axis*.

Militarily, we were weaklings. However, we had an advantage that the Axis powers didn't consider. It was the very thing they thought gave us a disadvantage. We were a democracy—a nation of free people. When free people set their minds to something, they become a powerful force. It took some time, but we became astonishingly strong.

Whether we wanted it or not, war was coming. We would win this war in our science laboratories and factories as well as on battlefields. The American people had been through a testing period that toughened them for a fight. The testing period was the Depression. We were used to tightening our belts and working hard. All of that, and more, was going to be necessary to win this war. It would be the most awful war in all of history.

29 Pearl Harbor

German tanks enter Czechoslovakia. Having to salute the invader was only the start of a conquered people's humiliation.

German troops arrive in Prague, March 15, 1939, the day Hitler announced, "Czechoslovakia has ceased to exist."

It is Sunday, December 7, 1941, and the sun is shining in Washington, D.C. To the morning churchgoers it seems just another bright winter day. At the White House, 31 guests are expected for lunch. There will be guests for dinner, too. None of that is unusual. The White House has become an informal, busy place since the Roosevelts moved in. That was more than eight years ago. FDR was reelected in 1936 and again in 1940. No other president has served more than two terms.

The American people (or most of them, anyway) have great faith in their president. These are dangerous times, and alarming things are happening all around the world. It is important to have a leader who can be trusted.

Hitler has steamrollered his way to some astounding victories. He has taken Austria, and Czechoslovakia, and Poland, and Finland, and Denmark, and Norway, and Holland, and Belgium. One by one he picked off all those countries. The democracies let him do it. The democratic nations are so sick of war that they are willing to do anything to try to avoid it. What they have actually done is to make the war much worse than it would have been if they had stopped Hitler earlier.

It was when the Nazis marched into Poland that Britain and France finally responded. (Both nations had pledged their help to Poland if it was attacked.) Britain and France went to war.

What they faced was something called *blitzkrieg*. That was the

The Polish army was hopelessly underequipped and outdated; what could cavalry with lances do against a tank?

The Nazis swept into Belgium and Holland simultaneously. Below, left, a German paratrooper enters Holland. The city of Rotterdam resisted so strongly that Hitler bombed it to pieces out of spite. Above, left, a Belgian family stumbles through air-raid rubble. Above, right, the incredible sea rescue on the beaches of Dunkirk.

German word for "lightning war," which was a good description. The Germans sped their troops, tanks, and artillery across nations, obliterating them almost before they knew what was happening.

When superbly trained, well-equipped German forces raced into France, the country was overwhelmed. A large British–French army was trapped at Dunkirk, on the English Channel. It looked as if the soldiers were doomed. Then the British government sent out an appeal for boats. Soon fishermen, dentists, grocers, tugboat captains—anyone with a boat that could make it across the Channel—were sailing, back and forth, back and forth, ferrying soldiers to England. They saved an army, but they couldn't save France. On June 14, 1940, German tanks rolled into Paris.

Now almost the only European democracy left is Britain. And Britain is under attack. German bombers are pounding that small island. It looks as if it will go next. Everyone knows that the Nazis plan to invade England. Hitler's goal is world conquest. Americans have plenty of reason to worry.

As if that isn't bad enough, the situation in East Asia is awful. Japan has conquered Manchuria and part of China, gone into French Indochina (now Vietnam, Laos, and Cambodia), and is threatening Thailand, the Philippine Islands, and some other places. The United States has sent letters to Japan objecting to this aggressive behavior. Inside Japan there is conflict: warlords are battling moderates for government power. The moderate Japanese premier asks to meet Roosevelt. The president doesn't understand the importance of the request. He refuses the meeting. The premier loses his job; he is replaced by a warlord.

This very day—December 7, when the sun shines so brightly in Washington—Secretary of State Cordell Hull receives a call from two Japanese diplomats. They ask for an emergency meeting. Hull expects to be given the Japanese government's answer to an American peace letter.

At the White House, after lunch, the president is working on his stamp collection. He began collecting stamps while he was still a boy. His good friend Harry Hopkins is in the sitting room with him; so is his Scottie dog, Fala. They are relaxing. The phone rings. It is close to 2 P.M., Eastern time.

Secretary of the Navy Frank Knox is on the line. His voice is quivering. A message has just been received from Hawaii. This is what it says: AIR RAID ON PEARL HARBOR—THIS IS NOT A DRILL.

Pearl Harbor, in the Hawaiian Islands, is where the Pacific Fleet is headquartered! On Sunday morning ships were lined up in the harbor; their crews were having breakfast, or relaxing, or sleeping. At 7:02 A.M. Hawaiian time, a radar operator saw some blips on his screen. The operator didn't pay attention to them. He thought they were bombers he was expecting from the West Coast.

By 7:55 A.M. he knew better. That was when the first dive bombers—with the red Japanese sun painted on their sides—let their bombs loose on Battleship Row. The battleship *Arizona* gave off a tremendous roar,

One week after France surrendered, Hitler was in Paris.

Japan's prime minister, Hideki Tojo, called the attack on Pearl Harbor "a blow for the liberation of Asia."

Far left, crewmen hold Japanese planes' propellers until they reach take-off revs. Naval lieutenant and pilot Zenji Abe said, "I was surprised at how quick American antiaircraft had responded [the spots in the sky, below, are antiaircraft flak]. Over Pearl Harbor there was black smoke. I tried to find my target. I chose a big ship. That was the *Arizona* [left, after the attack], I found out later."

ALEUTIANS

Departure Nov. 26.

JAPAN

GUAM CITY

•WAKE I.

MIDWAY

HAWAII

Launching Point

PEARL HARBOR

103 HIGH LEVEL BOMBERS
132 DIVE BOMBERS
40 TORPEDO BOMBERS
79 FIGHTERS

SHIPS SUNK or SEVERELY DAMAGED
① ARIZONA
② TENNESSEE
③ MARYLAND
④ CALIFORNIA
⑤ NEVADA
⑥ VESTAL
⑦ W. VIRGINIA
⑧ OKLAHOMA
⑨ HELENA
⑩ OGLALA
⑪ SHAW
⑫ PENNSYLVANIA
⑬ CASSIN
⑭ DOWNES
⑮ HONOLULU
⑯ RALEIGH
⑰ UTAH
⑱ CURTISS

FORD ISLAND

BATTLESHIP ROW

SUBMARINE BASE

U.S. NAVAL STATION
PEARL HARBOR

MINESWEEPERS

OIL STORAGE TANKS

split in two, and slipped to the bottom of the harbor. That was just the beginning. All the American planes on the island were damaged or destroyed. Most of the warships were crippled or sunk. And more than 2,000 soldiers, sailors, and civilians were killed.

At 2:05 P.M. Washington time (which is 8:05 A.M. Hawaiian time), the Japanese envoys arrive at Secretary of State Cordell Hull's door. They are part of an elaborate Japanese plan of deception, but their timing is off. Before the secretary can see them, his phone rings. It is the president, with the awful news of the Japanese attack. Now, Hull is from Tennessee, and he claims he has a Tennessee temper. The stories of what he says to those envoys will differ, but it is known that they leave quickly, with their heads down.

Hull is soon at the White House. So are many government and military officials. Newspaper reporters begin arriving. At 2:25 P.M. the story goes out on news wires to the American people. The reports from Pearl Harbor are humiliating, but that isn't the only bad news. This same day, the Japanese have attacked American and British bases at Midway, Wake Island, Guam, Hong Kong, Singapore, and the Philippines.

It is an astonishing act of aggression. But this president is at his best in a crisis. His advisers are angry, fearful, and frustrated. The president remains calm. He came into office during the nation's worst economic crisis. This is worse: the free world is fighting for survival.

Secretary Hull with Japanese ambassador Nomura (left) and special envoy Kurusu on their way to the White House, three weeks before the bombs fall on Pearl Harbor.

Pearl Harbor is a disaster, but it may also be a lucky break. It unites the nation. There are no more isolationists. Everyone joins the war effort. Pearl Harbor shows the damage that air power can do. It changes people's thinking on that subject.

The next day the president goes before Congress. The Japanese have launched an "unprovoked and dastardly attack," he says. December 7 is "a date which will live in infamy." He asks Congress to declare war on Japan. Three days later, Japan's allies—Germany and Italy—declare war on the United States. It is World War II. It will make the awful First World War seem like a fire drill. The United States will fight this war against the wolves, maintain its democracy (as it did during the terrible Depression), and remain, as Abraham Lincoln said, the last best hope of earth.

Infamy (IN-fuh-me): it means *evil reputation*.

January 1942: farewell from workers in Brooklyn's Navy Yard, who have volunteered to go out to Hawaii to help rebuild the shattered fleet.

The True Goal We Seek

Franklin Delano Roosevelt knew how to talk to people. He knew how to explain complicated things in simple language. As soon as he became president, he decided to use the radio to explain what the government was going to do about the Depression. He was as relaxed and informal as a friend sitting in the living room. He called his radio broadcast a "fireside chat." Soon those fireside chats became a regular thing. Americans liked listening to their president, especially since he had a good sense of humor. But on February 9, 1942, he didn't have anything funny to say. This is part of his speech to the American people that day:

We are now in this war. We are all in it—all the way. Every single man, woman, and child is a partner in the most tremendous undertaking of our American history....On the road ahead there lies hard work—grueling work—day and night, every hour and every minute. I was about to add that ahead there lies sacrifice for all of us. But it is not correct to use that word. The United States does not consider it a sacrifice to do all one can, to give one's best to our nation, when the nation is fighting for its existence and its future life....There is no such thing as security for any nation—or any individual—in a world ruled by the principles of gangsterism. There is no such thing as impregnable defense against powerful aggressors who sneak up in the dark and strike without warning. We have learned that our ocean-girt hemisphere is not immune from severe attack—that we cannot measure our safety in terms of miles on any map anymore....

The true goal we seek is far above and beyond the ugly field of battle. When we resort to force, as now we must, we are determined that this force shall be directed toward the ultimate good as well as against immediate evil. *We Americans are not destroyers; we are builders.*

We are now in the midst of a war, not for conquest, not for vengeance, but for a world in which this nation, and all that this nation represents, will be safe for our children. ...We are going to win the war and we are going to win the peace that follows.

And in the dark hours of this day—and through dark days that may be yet to come—we will know that the vast majority of the members of the human race are on our side. Many of them are fighting with us. All of them are praying for us. For, in representing our case, we represent theirs as well—our hope and their hope for liberty under God.

30 Taking Sides

After their first meeting, Churchill got a cable from FDR: "It is fun to be in the same decade with you."

FDR smoked cigarettes through an extra-long cigarette holder, and Winston Churchill puffed on an ever-present cigar. Few people then knew that smoking shortens lives and increases the risk of disease.

This is how the war was fought:

On one side was the Berlin–Rome–Tokyo Axis led by Adolf Hitler, Benito Mussolini, and Japan's premier, General Hideki Tojo.

On the other side were three major Allied forces. Who were they? Who were their leaders?

President Franklin Delano Roosevelt represented the United States. His was already a voice of freedom respected all over the world.

Britain, the second Allied power, was led by a man with a pudgy baby face. His name was Winston Churchill, and, like Roosevelt, he had a powerful voice and was inspiring when he spoke. Churchill had an American mother and a father who was a British lord. When he was young he was a poor student, but he got better. He went to Sandhurst—the British West Point—and became an army officer, a good one. He got medals for bravery. Then he became a newspaper reporter, learned to pilot a plane, wrote history books, entered politics, and became a member of Britain's Parliament. He was one of the first Englishmen to see the danger of Hitler's Nazi Party and to speak against it. That was when most people in England and America were acting like ostriches. They buried

The Russians were stunned when Hitler invaded. The Nazi–Soviet Pact was dead, and they had a new role as Allies.

Berlin, Rome, and Tokyo are the capitals of which nations?

their heads in the sand, closed their ears, and didn't want to hear anything about war.

The third Allied power? Was it France? No. France was under German control. (However, as you know, a free French army was saved at Dunkirk. It was led by General Charles de Gaulle, and it did fight with the Allies.)

How about China? Was China the third power?

No. China was fighting Japan, although many Chinese were unhappy with their leaders. Civil war was brewing in China. China was in turmoil.

You may have a hard time believing what the third Allied power was, but here it is: Soviet Russia (the U.S.S.R.). Russia's leader, in 1941, was dictator Joseph Stalin. He was head of the Soviet Communist Party. He ruled Russia using secret police and terror. Many of his own people hated him. But others were fooled by Stalin. Roosevelt may have been one of them. Dictators often have charm, and Joe Stalin had a lot of charm—when he wanted. But what mattered was that he was fighting Hitler and so were we. As Winston Churchill said, "If Hitler invaded Hell I would make at least a favorable reference to the Devil in the House of Commons."

Russia, however, didn't start out on the Allied team. Here is some background.

In 1939, Hitler and Stalin signed a friendship pact. They said they would not fight each other. They made plans together to march into Poland—one from the east, the other from the west—and to gobble up that nation. They did it. Poland was squashed and divided.

Before the Nazi army marched into Poland, Hitler told his generals:

The victor will not be asked afterward whether or not he told the truth. In starting and waging war it is not right that matters but victory. Close your hearts to pity! Act brutally!...The stronger is in the right.

His generals did as they were told.

That was when France and England finally realized that they couldn't avoid war—although now it would be a difficult one. They had let Germany build a huge military force. At the same time, the free nations had cut their armies and navies. There wasn't a lot anyone could do when Germany marched armies into Belgium, Holland, Luxembourg, and France. The Nazis were winning everywhere.

The German air force—the Luftwaffe (LOOFT-vah-fuh)—soon began dropping bombs on England—tons and tons of bombs.

A cartoon sums up the Allied view of Stalin—cozying up to Hitler and his chief aide, Göring—before Germany turned the tables, double-crossed Stalin, and changed the whole course of the war.

It is hard to lead a country in exile, and many said Charles de Gaulle was arrogant—including FDR, who couldn't stand him and said he was "a nut." But he was a good general and kept underground resistance going in France throughout the German occupation.

In August 1940, the first German bombs fell on London. Night bombing of London—it was known as "the Blitz"—began in September and went on until June. The Londoners stuck it out.

Then Hitler made some stupid moves. First, he went into Greece and Yugoslavia, where his forces faced some heroic fighters. That stopped Germany for a while. Then Hitler doublecrossed Stalin. He decided to invade Russia. That had been his plan all along. He had even written a book, called *Mein Kampf* (it means "my struggle" in German), that told all about his goal of world domination. "No human being has ever declared or recorded what he wanted to do more often than me," he said.

Anyone who read Hitler's writings knew what he had planned. He said that Germans were the "master race," and that they needed more room, and other nations to serve them as slaves. Churchill and Roosevelt paid attention. They knew Hitler was capable and effective, as well as evil. For a long time, most other politicians just didn't take him seriously.

But Hitler didn't intend to share power. The Russians were a threat to his goal of world domination. So, when some of his generals told him not to go into Russia, he didn't listen. He needed oil and wheat and other resources from Russia. Besides, he thought Russia would be an easy victim. So did experts everywhere. The American secretary of war predicted that it would take Germany three months to conquer Russia.

The damage done in London by the fires the bombs caused was as bad as their explosive effect. The great cathedral of St. Paul's, seen here during the worst raid of the Blitz, survived intact.

Look at a map. The Germans prepared the most massive army ever assembled. Their forces stretched from Finland to the Black Sea. They attacked with the latest in military equipment: tanks, bombs, and artillery.

At first, the Germans had an easy time of it. The Russians weren't prepared; much of their military equipment was out of date. Hitler instructed the German army to turn Russia into a slave nation. The Nazis murdered millions of Russians.

Now look at a map again. Notice the size of Russia. Look at the latitude of cities like Moscow and Leningrad (which today is called by its old name, St. Petersburg). They are cold places. Winter came early in the fall of 1941. The first snow fell in Moscow on October 2. The Germans weren't prepared for the cold. They got stuck in a Russian

Left: children outside what is left of their London home. Many city kids like these were "evacuated"—taken in by families in the country. Their white bags hold gas masks.

The Second World War was a truly global conflict, involving several continents and many nations and peoples (unlike World War I, which was fought largely in Europe and around the Mediterranean Sea).

Russia's winter had stopped Napoleon, too. Hitler should have remembered that.

131

Even the cleated caterpillar tracks of the German tanks (above) bogged down in the snow. One of the victors in Russia was winter.

The Russian city of Leningrad was almost totally cut off from September 1941 until January 1943. People had no electricity or heat; water was contaminated because mains burst (below, collecting water from a broken main). Left, a small shrouded body on a sled.

winter. It happened to be the coldest winter in 30 years. The temperature dropped to 60° below zero. The German army was far from home, and having a hard time getting supplies and food. German soldiers froze. German soldiers starved. So did Russians. It was a disaster for both sides. But the Russians were at home, and able to outlast the hungry, discouraged German army.

FDR sent his friend Harry Hopkins to Russia. "Give us antiaircraft guns and aluminum and we can fight for three or four years," said Stalin. He was right. We sent guns, aluminum, food, tanks, planes, and more. The Russians gave their lives.

The Russians went all out in their fight against Nazi Germany. No nation fought any harder.

No one knows how many Russians died in World War II. Some say 15 million. Some say more. No other country has ever suffered such war losses. Russia

Leningraders ate dogs and rats and Vaseline and made soup from the glue in furniture or wallpaper. Many died of hunger anyway.

was our ally and friend during the world war. But Russia under Soviet communism was a dreadful place. Stalin was a vicious dictator. Stalin expected something for fighting Hitler. What he expected, and got, was domination over the other countries of eastern Europe. Could Hitler have been destroyed without Stalin's help? Perhaps not. Certainly it would have taken many, many more American and British lives.

31 World War

Loading machine-gun cartridge belts for dive bombers at Norfolk, Virginia.

Billy Mitchell was right: air power changed war. In World War II more people were killed by bombs or pieces of shells (called *shrapnel*) than by bullets. In World War II cities were bombed; huge civilian populations were massacred.

There was something else about air war: it made killing a mechanical act. Imagine being in the infantry. You see the enemy eye to eye; it makes you realize the enemy is just like you—human. Officers know that some soldiers are never able to pull their triggers. They are never able to murder—even to save themselves. But a bomber pilot doesn't see his victims. A bomb can't tell the difference between an enemy soldier and a child on her way to school. It will kill them both.

An enormous number of bombs were dropped—by both sides—during World War II. Billy Mitchell thought air power would eliminate the need for foot soldiers. He was wrong about that. There was still plenty of old-fashioned infantry fighting.

Look at the world map on page 131. World War II was truly a world war. Here are just a few of the places where American troops fought; see if you can find them in an atlas.

France, Germany, Tunisia, Sicily, Italy, Morocco, Burma, Guam, Malaysia, Philippine Islands, Wake Island

Now imagine you are a general and you are planning a battle on a Pacific island. Suppose you want to get 15,000 men onto the island

Manning a sub's periscope.

The U.S. drafted black men but segregated them and often assigned them to service jobs instead of combat units. Below, a black fighter squadron.

Above left, paratroopers of the U.S. 82nd Airborne Division drop over Belgium in 1944. Right, one method of getting soldiers—along with their trucks and all their ammunition and supplies—across a river is with a kind of barge. This one in Burma is powered by ordinary outboard motors.

U-boat is the abbreviation for the German word *Unterseeboot*—"undersea boat."

A **torpedo** is like a giant bullet with a propeller that travels through water and can sink a ship.

and surprise the enemy. How are you going to do it?

A parachute drop?

Maybe, but remember, parachutes make great targets. You'd be better off bringing them in by boat. Many of those islands don't have deep harbors, though. Big ships can't come in close.

Can the soldiers swim in?

Not with their guns and artillery and trucks and tanks and food and ammunition and medical supplies.

We're going to have to invent and develop new kinds of landing equipment and war gear. And we're going to have to do it very fast. We'll design huge landing craft that have big rooms—called *holds*—that can be flooded to form miniature lakes so that boats can zoom out. We'll design other landing ships that will carry tanks and trucks as well as men. We'll design *amphibious* (am-FIB-ee-us) vehicles that will go on land or water. One of the most useful—a truck that swims—will be called a *duck*. Another new, tough vehicle—which can handle rough roads, mountain passes, and rutted fields—we'll call a *jeep*.

We'll design superb submarines that can stay under water for months at a time. Then we'll design torpedoes and depth charges to destroy submarines. The Axis nations will be doing the same thing. Submarine warfare will be very important in this war. German subs are called *U-boats*, and the Atlantic Ocean is full of them.

We're going to do amazing things in medical science so that disease and infection will no longer be the major causes of wartime deaths. The lives of many badly wounded men will be saved.

All through the war we will keep improving our weapons, planes, tanks, and armored vehicles. The Germans and Japanese have a head start on us. They have fine scientists and technicians. This war will be-

come a race to see who can produce the best weapons fastest. The Germans are working on rockets—called V-1s and V-2s—that are devastating. Luckily, it will take most of the war to get them perfected. When they start shooting rockets at England there will be many, many deaths. (The V-2 rockets are being designed to hit the United States.) We are behind on rocket development. After the war, German rocket engineers will tell us they got many of their ideas by studying the work of our rocket expert Robert Goddard.

We know something that they don't suspect we know. They think they are smarter than we are. They are wrong. We have learned to read their most difficult codes. That will prove more valuable than almost anything else we do.

Have you ever tried writing in code? It's easy. Just put numbers in place of letters and you have a code. Armies have always needed codes. Suppose a general wants to tell a faraway commander to attack. He sends a messenger. But he wants his orders in code in case the messenger is caught. He certainly doesn't want the enemy to know his plans.

In George Washington's day, a screen was sometimes put over a piece of paper. There were holes in the screen. The secret message was the words that showed through the holes. Everything else was there to fool you.

During World War II, both sides moved huge armies and navies and tried to do it secretly. Most orders were sent by telegraph. Anyone could listen. So codes were vital. They became very complicated. The Germans

Getting from A to B: top left, lowering a jeep from a Coast Guard assault transport into a landing craft. Top right: each of these tiny cars carries a real bomb. Above, an aerial photo of U.S. troops wading ashore from landing craft onto Morotai Island, between New Guinea and the Philippines.

A V-1 bomber over London.

and Japanese thought no one could possibly figure out their complex codes.

We cracked the Japanese secret code even before the war began. Solving the German military code was much harder. German coded messages were sent and received on special machines. Then a German tank was captured in Poland. It had a code machine inside. The machine was smuggled out of Poland to England. When it got to England no one could figure out how to work it. The English called the code machine "Enigma." An *enigma* is a puzzle. They put some of their best scientific and mathematical minds on the job of solving the puzzle. It was incredibly difficult. How they did it is a fascinating story. Several books have been written about it. You can find them in the library.

Once the code was broken, we knew almost everything the Axis powers were planning to do. Now we Allies had to pretend that we didn't know some things. We didn't want the codes to be changed.

Germany's racial policies have caused many of its best scientists to flee the country. That helps us and slows their progress.

When American naval forces capture a German submarine off the coast of West Africa, they find a newly developed torpedo and a secret radio code on board. They pretend they have sunk the sub, so the Germans won't change the code.

Cryptography Means Codemaking

In World War II, codemakers (who all seemed to be geniuses) created extraordinary code machines in order to write secret languages that would baffle the enemy. But codebreakers were, if anything, even smarter than the codemakers. Just about all the codes did get broken—except for one that baffled all the geniuses. No one could figure it out. Maybe that's because it happened to be a real language, spoken by real people, who were faster than any of the fancy machines.

The language was Navajo, and it was spoken by 420 marines who called themselves *Dineh*—the People. In western movies, Indians are usually known for their silence. These Native Americans did plenty of talking. They made up their own code using their own words: Hitler was *Daghailchiih* (mustache smeller), bombers were *jaysho* (buzzards), and bombs, *ayeshi* (eggs). Navajos landed on every major island in the Pacific. Major Howard Conner said, "Without the Navajos the marines would never have taken Iwo Jima." They were a secret weapon in the Pacific.

Enigma. The possible number of encoding positions for each letter was unbelievably huge: 5,000 billion trillion trillion trillion trillion.

32 A Two-Front War

Tarawa, in the Gilbert Islands, where marines put up this lonely signpost, was a tiny spot in the Pacific. Yet it took bitter fighting to capture it.

Did you ever hear of Janus, the two-faced god of Roman mythology? *January* was named for Janus; he was the god of doorways and gates who looked in two directions at the same time.

During World War II the United States was like Janus. We had to look in two directions at the same time. We were fighting a two-ocean war. That was a terrible problem for our generals and admirals. How do you divide your forces? How do you protect two huge coasts from attack?

Looking west (to East Asia), the view was awful. The Pacific theater was mostly a disaster. (Military officers call a war region a *theater*. Strange, but that's the way it is.) The Japanese moved like lightning. Their forces were well trained and well equipped. At the start, Japan seemed to do everything brilliantly. Remember, on December 7 they didn't just bomb Pearl Harbor—they attacked a whole string of strategic spots, almost simultaneously! In just a few months the Japanese captured Thailand, the Philippine Islands, the Malay peninsula, Java, Burma, Guam, Wake Island, the Gilbert Islands, Singapore, and Hong Kong.

Simultaneously (sy-mul-TAY-nee-us-lee) means "at the same time."

In the first six months of war, under General Tojo (left), Japanese forces such as the imperial marines (right) overran more territory than any conqueror since Napoleon.

In May 1942, in the battle of the Coral Sea, both the U.S. and Japan suffered heavy losses: here the crew of the sinking aircraft carrier U.S.S. *Lexington* bail out frantically. But the Japanese lost two of their carriers and had to retreat.

Black and White Blood

Charles R. Drew was uncommonly gifted. He was a star athlete (in football, basketball, baseball, and track!), a brilliant student at Amherst College, an outstanding doctor (he was a professor of medicine at Howard University), and the man who developed the idea of a blood bank for storing blood plasma (during World War II). In 1942, he organized the blood-bank programs for both the U.S. and Britain, and supervised the Red Cross's blood-donor program. Besides all that, he had an amiable personality—people liked him.

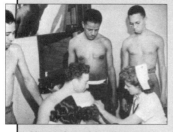

Home-front blood donors, 1944. The black men's blood will be segregated.

But he was enraged that black blood and white blood were segregated in blood banks, and he spoke out against that absurdity. He was killed in a car accident. A myth has grown that because of his color, he was refused treatment at an all-white hospital. That is not true. What is true is that his talents were often frustrated by the idiocy of prejudice.

Check those places on a map and you'll see: the Japanese controlled East Asia. People in India and Australia were trembling. They thought they were next.

The Western theater—Europe—wasn't any better. As you know, the Nazis controlled most of Europe. They even had troops in North Africa; the Mediterranean was a kind of Nazi sea.

So was the Atlantic Ocean. That was because of the German U-boats. England and Russia were desperate for help. The United States had to ship weapons, tanks, oil, and men across the Atlantic to Europe. But the U-boats seemed to be everywhere. In the first four months of 1942, almost 200 of our ships were sunk. One summer day, people in Virginia Beach, Virginia, watched in horror as a ship was torpedoed and sunk in sight of the beach. Ships were being sunk faster than they could be built. Each time a ship was torpedoed, American men drowned.

Somehow, people in this country didn't get discouraged. We were convinced we could win this war and we set about doing it. Dr. New Deal turned into Dr. Win the War. He became a great war president. No matter how gloomy things seemed, President Roosevelt remained confident and optimistic. He gave courage to the nation.

He had good people to work with. General George C. Marshall, his chief of staff, was a superb general, and modest. Someone said he had the wisdom of George Washington and the strategic sense of Robert E. Lee.

Tough, experienced Admiral Ernest J. King, chief of naval operations, did his job well. So did admirals Nimitz and Spruance and generals Dwight D. Eisenhower, H. H. "Hap"

View from a U-boat before and after an attack: right, looking through the periscope; above, a torpedo has split the freighter in two.

Arnold, Douglas MacArthur, and others. Remember all the trouble Abraham Lincoln had with his generals? Roosevelt was lucky; the nation was lucky too.

But the first battles were grim. Admiral King warned, "The way to victory is long; the going will be hard." He was right. We started out as losers. Then things began to change. Maybe it was because the Japanese got greedy. They didn't know when to stop. They thought they were invincible—which means unbeatable. Nobody is unbeatable. We had been taking a pounding in the

Dwight D. Eisenhower

Douglas MacArthur

Right: the 165th Infantry reaches Makin atoll in 1943 (Tarawa was captured at the same time). The coral bottom makes for very hard wading. Above, a water buffalo assault vessel near Guam.

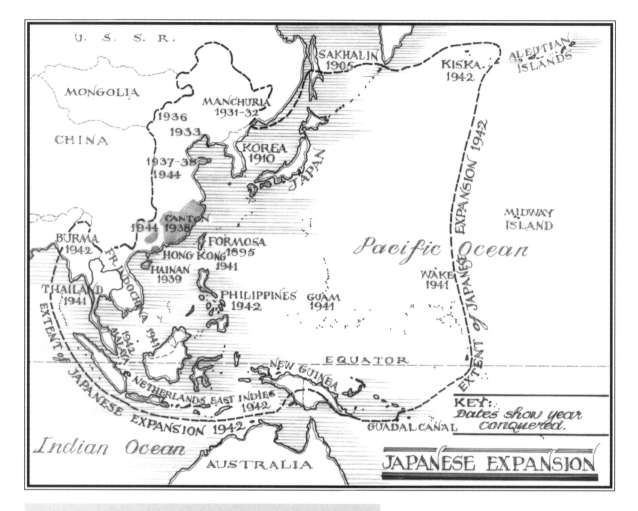

Map labels: U.S.S.R.; MONGOLIA; MANCHURIA 1931-32; CHINA; 1936; 1933; KOREA 1910; 1937-38 1944; JAPAN; SAKHALIN 1905; KISKA 1942; ALEUTIAN ISLANDS; MIDWAY ISLAND; Pacific Ocean; CANTON 1944 1938; BURMA 1942; FORMOSA 1895; HONG KONG 1941; HAINAN 1939; FR. INDOCHINA 1941; THAILAND 1941; PHILIPPINES 1942; GUAM 1941; WAKE 1941; EXTENT OF JAPANESE EXPANSION 1942; MALAYA 1942; NETHERLANDS EAST INDIES 1942; NEW GUINEA; EQUATOR; GUADALCANAL; EXTENT OF JAPANESE EXPANSION 1942; Indian Ocean; AUSTRALIA; KEY: Dates show year conquered.; JAPANESE EXPANSION

1943: An American PT (propeller-torpedo) boat is rammed by a Japanese destroyer. Two Americans are killed. Eleven others cling to the boat and then swim to a nearby island. Their commander carves a message in a coconut and gives it to friendly islanders, who bring help. The commander's name is John F. Kennedy.

Pacific. But when we won three big victories, the Japanese learned to respect our fighting ability. The victories were in the Coral Sea, at Midway Island, and at Guadalcanal.

The Coral Sea battle wasn't exactly a victory. There was a series of mistakes on both sides—big mistakes. Our losses were heavier than the enemy's. But we prevented them from capturing a strategic base in New Guinea, and that may have saved Australia from invasion.

The Japanese expected an easy win at tiny Midway Island. If they controlled that island they would control the air over Hawaii. Then, perhaps, they could attack our West Coast. In addition, they intended to destroy the ships in our fleet that had escaped Pearl Harbor—especially our aircraft carriers. Their plan was to lure us into a trap and surprise us. What they didn't know was that we could decode their secret messages. We knew of their plans. Still, they almost won. The battle over Midway was ferocious.

At first, things were awful for us. Forty-one of our torpedo bombers took off after the Japanese ships, and most were shot down. Then our dive bombers arrived, and destroyed four massive Japanese aircraft carriers. It was the first defeat the Japanese navy had suffered. And it was a battle that proved that air power would be the key to this war.

The battle for Guadalcanal was something else. If you hold on for a chapter I'll tell you about it. But before that you need to learn about something going on at home.

The battle of Midway took two days. U.S. Navy bombers such as the one below—the Japanese called them "hell divers" —caught the Japanese in the process of refueling, and sank a number of ships, including the cruiser *Mikuma* (bottom).

33 Forgetting the Constitution

The American citizen who owned this store put up the sign the day after Pearl Harbor was bombed. Like so many others, he had to sell his business at a sacrifice price and go to a camp.

Haruko Obata lived in a house in lovely, tree-shaded Berkeley, California. Her father was a professor at the University of California. Haruko was an American citizen. Like most Japanese Americans, she was proud of her Asian heritage, but she didn't approve of the ways of the warlords who ruled the Japanese empire. In school she studied the Constitution and its guarantees. She was happy to live in the land of liberty.

Then, one day, Haruko's world changed. Her father came home and told the family that they were moving. They had just a few days to get ready. They could take only as much as they could carry. They might never again see the things they would leave behind. They were going to live—against their wishes—in a prisonlike camp.

What had they done? Just a minute, and I'll get to that. But first, imagine that you are Haruko. You have some hard decisions to make, and you need to make them quickly. What will you

The Japanese internment camps on the West Coast, and the internment camp for Jews in Oswego, New York, were run by the same government agency. Even the food was the same.

One girl was seven when her family was taken to a camp in the California desert. "Someone tied a numbered tag to my collar and to the duffel bag," she wrote. Nobody explained why they suddenly had to move.

The only places with enough room to house so many people are too cold, hot, or windy, places no one wants to live in—swamps, mountains, or deserts, like Manzanar (above).

choose to take with you? Sorry, your dog can't go. You'll have to give her away. Books? Games? Toys? Not if they are heavy. No one knows exactly where this camp is. It may be very cold. Or hot. You won't be able to take much besides clothes. Your parents must sell the car, the house, and almost all their possessions; because they do it so quickly, they will get hardly anything for them.

You and your family are going to a camp that is surrounded by a barbed-wire fence. Armed guards stand in watchtowers. If someone tries to walk out into the desert he will be shot. What have you done that is so terrible? Why are you and your family in this prison camp?

You have not done anything wrong. Yes, you read that right. The Obatas have done nothing at all. They have been fine citizens.

But they are of Japanese descent, and the United States is at war with Japan. There is anti-Japanese hysteria in America, especially in California. Some of it is understandable. War is terrible. The Japanese government is horrible. But the Japanese in America have nothing to do with that. Some people don't understand that. Many authorities expect the Japanese to attack the West Coast. A Japanese submarine actually fires shells that land—harmlessly—in the Los Angeles area.

The whole evacuation operation was so rushed that very few camps were anywhere near completion when the inmates arrived. At Tanforan (above), where Haruko Obata was sent, the center had been open for only two days when the photograph was taken, and people had to line up for hours just to get fed.

Internees registering at Santa Anita, California. Then everyone had to open their baggage for inspection before being bused to a camp.

143

A young woman named Miné Okubo was interned at Tanforan and then in Topaz, Utah. She drew pictures for a book about her life called *Citizen 13660*. The clothes the camps supplied never fit; the Utah mosquitoes were fierce; and the walls were so thin and badly built that quiet and privacy were impossible.

People are terrified. There are rumors that Japanese-American fishermen are sending signals to Japanese ships and planes. None of that is true, but in wartime, how can anyone be sure?

Most Japanese Americans feel anguish. They love the United States, its opportunities and its inspiring vision. But they also take pride in their ancient Japanese heritage. For them, this war is like a civil war.

In addition, the Japanese in America face a special problem. It is an old problem. It is racism. A racist law prevents Japanese immigrants from becoming citizens. However, anyone born here is automatically an American citizen. Two-thirds of the Japanese Americans are *Nisei*—the Japanese word for those born in America. The war provides an opportunity for the racists, and for the greedy, too. They can get rid of the Japanese Americans and grab their land and possessions. They can scream and accuse, and, because of the war, some people—some very good people—will believe the lies they tell.

An army general who is in charge of the defense of the West Coast makes up stories of treason and treachery in the Japanese community. His superiors believe him. There is no reason for them to do so. The general has no evidence. But the leaders of the War Department don't take time to investigate. They are fighting a war.

The attorney general speaks up. He reminds the secretary of war that the Fourth Amendment to the Constitution protects people from "unreasonable searches and seizures." The 14th Amendment says "nor shall any State deprive any person of life, liberty, or property, without due process of law." But in the War Department they are worried about wartime "security." They believe it more important than the Bill of Rights and its guarantees. They are not concerned about habeas corpus. They urge the president to put all Japanese Americans in prison camps until the end of the war. They say it is necessary for the safety of the nation.

President Roosevelt issues an executive order. The 120,000 Japanese Americans are given a few days to get ready. They will be sent to 10 different internment camps.

Here is how Haruko Obata described her arrival at the Tanforan camp:

> *When we arrived at Tanforan it was raining; it was so sad and depressing. The roadway was all mud, thick mud and our shoes would get stuck in mud when you walked outside.*

They gave us a horse stable the size of our dining room with a divided door where the horse put his head out—that was our sleeping quarters. There were two twin beds made of wood, bunk beds, and another bed on the opposite wall. It was supposed to be a couch but it was made of wood too. There was nothing else. Nothing. That one time I cried so much. That was the only time I cried; it was awful.

There is much more to this story, much, much more. Mostly it is of a people who—as soon as they got settled—didn't cry. They did their best in a bad situation. They planted seeds and grew crops. They raised farm animals. They fed themselves and sent their surplus to support the war effort. They fixed up their sleeping quarters. They established schools, churches, recreational centers, newspapers, scout troops, baseball teams, and their own camp governments.

Some were let out of the camps to work in war factories. Many became soldiers. A Nisei regiment fighting in Europe won more commendations than any other regiment in the whole United States Army. Infantryman Harry Takagi explained:

We were fighting for the rights of all Japanese-Americans. We set out to break every record in the army. If we failed, it would reflect discredit on all Japanese-Americans. We could not let that happen.

More than 16,000 Nisei served in the Pacific, most in military intelligence work as interpreters for the army and navy. Some went behind enemy lines as American spies. Japanese-American women volunteered and served in the Woman's Army Corps, as army nurses, and in the Red Cross.

At first, people in the War Department objected to the idea of Nisei serving in the army. But, finally, President Roosevelt spoke up. He said:

The principle on which this country was founded and by which it has always been governed is that Americanism is a matter of the mind and heart; Americanism is not, and never was, a matter of race or ancestry.

Japanese-American soldiers take cover from a German shell in Italy.

If you don't know what habeas corpus is, see Book 3 of **A History of US.**

They issued us army mess kits, the round metal kind that fold over, and plopped in scoops of canned Vienna sausage, canned string beans, steamed rice that had been cooked too long, and on top of the rice a serving of canned apricots.... Among the Japanese, of course, rice is never eaten with sweet foods, only with salty or savory foods. Few of us could eat such a mixture. But at this point no one dared protest. It would have been impolite. I was horrified when I saw the apricot syrup seeping through my little mound of rice. I opened my mouth to complain. My mother jabbed me in the back to keep quiet. We moved on through the line and joined the others squatting in the lee of half-raised walls, dabbing courteously at what was, for almost everyone there, an inedible concoction.

—Jeanne Wakatsuki Houston, *Farewell to Manzanar*

145

The Hirano family with a photograph of their other son in uniform. They were sent to Arizona.

Los Angeles photographer Toyo Miyatake and his family were sent to Manzanar. He wasn't allowed a camera, so he made one and took secret pictures of life behind the barbed wire.

In the course of the war, ten people were convicted of spying for Japan. All were white. Not a single Japanese American was ever accused of helping the enemy.

Eventually, the camps were closed and people went out and did their best to build new lives. It wasn't easy; they had lost all their possessions. Many still faced racism when they tried to find jobs and new homes.

Now, you may be thinking that racism is terrible, but that, really, it is only a problem for the people who are made to suffer. Don't be fooled. Hatred is a contagious disease; it spreads quickly. It is like the poison gas that both sides used during the First World War; when the wind changed, it blew back on the gassers.

No people is immune to the virus of hate. It is how they handle it that decides the kind of people they are. This nation was founded on the idea that *all men are created equal*. A group of women at Seneca Falls, New York, changed that to *all men and women are created equal*. That is what we believe. That is what this nation is all about.

> *We the People of the United States, in Order to form a more perfect Union, establish Justice, insure domestic Tranquillity, provide for the common defense, promote the general Welfare, and secure the Blessings of Liberty to ourselves and our Posterity, do ordain and establish this Constitution for the United States of America.*

Remember George Washington's words:

> *The government of the United States…gives to bigotry no sanction, to persecution no assistance.*

If the hate virus comes creeping into your neighborhood, make sure you recognize it. Don't let yourself get infected. The more you read history, the more you will realize that the haters never win. Eventually, they get found out and put down. Often, unfortunately, it takes time.

Forty years after the end of the war, the American government officially apologized to the Japanese Americans for the terrible injustice done to them during World War II. Those who had been in the camps were given money in partial payment for their suffering. Today, when we Americans think back on the internment camps, we feel shame.

34 A Hot Island

Admiral Ernest King was a stern, opinionated man who was hardly ever known to smile.

General Eisenhower, who was a likable fellow, wrote in his diary in March of 1942: "One thing that might help win this war is to get someone to shoot King."

He was kidding. The King he was talking about was Admiral King. They disagreed on strategy. Do you think wars are easy to plan? Do you think the leaders all agree on how to go about it? Not often.

Most of our military leaders believed we should fight the war in Europe first and then the war in the Pacific. That made sense. Splitting your fighting forces is never a good idea. Besides, we didn't yet have enough supplies for two regions. But Admiral King said we couldn't just sit back and let the Japanese take over the Pacific. If we did, they would become so powerful that it would be almost impossible to win the war against them.

When the Japanese started building an airfield on an obscure island in the Solomon Island chain, Admiral King said that the United States needed to go on the offensive. So far—in Europe and the Pacific—we had been defensive fighters. King insisted that we take that island from the Japanese. It was an

U.S. troops of the 160th Infantry Regiment going ashore from a landing boat at Guadalcanal. It seemed like paradise—until they got past the beach.

Do you remember when Thomas Jefferson wrote to James Madison about Patrick Henry? He said, "What we have to do, I think, is devotedly pray for his death." Patrick Henry was fighting some of Jefferson's ideas. Do you think Jefferson really wanted him dead?

Jungle warfare can be as much of a fight with the jungle as with the enemy. Some men's clothing was damp so long, it rotted on their bodies.

U.S. Marine Raiders and their dogs, which were used for scouting and sending messages. One platoon had specially trained dogs that tracked down Japanese hiding deep in the jungle.

important decision. Not everyone agreed with it. It would cost many lives—American and Japanese. It turned out to be a decision that helped win the war.

The obscure island was named Guadalcanal, and it was such an out-of-the-way place that no one even had a map of it. But it was the right spot for a war base.

Find Australia on a map. Then look north, to New Guinea. To the east of New Guinea are the Solomon Islands. Guadalcanal is one of the southernmost of the Solomons. Anyone who has an airbase on Guadalcanal can make big trouble for ships and airplanes going to Australia, New Zealand, or even Japan (which was where the American military planned to go eventually). We couldn't let the Japanese put planes on that island.

From the air, Guadalcanal looks like a heavenly place: very green, with high mountains, thick forests, and jungles filled with wild orchids and bright-feathered and beaked tropical birds. To that picture, add sandy beaches, coconut palms, and banana trees. Does it sound like a place you'd like to visit? Well, the men who fought there called it "a bloody, stinking hole."

Guadalcanal is intensely hot (note the nearness of the Equator). Its jungles are filled with monster leeches, huge scorpions, poisonous centipedes, giant ants, writhing snakes, skulking rats, snapping crocodiles, and hungry anopheles (uh-NOF-fuh-lees) mosquitoes (whose bites bring malaria).

During the Spanish–American War, Walter Reed, an American doctor, discovered that quinine (KWY-nine) cures malaria. Quinine comes from a plant found in Java. The Japanese had captured Java. Doctors were working on synthetic quinine, but not fast enough for the troops who fought on Guadalcanal.

Most of Guadalcanal is tropical rainforest—which means steamy, thick jungle, a whole lot of rain, and black, squishy mud that comes up to a man's knees. Where there isn't rainforest there is kunai (KOON-i) grass. Kunai grass blades are saw-toothed, stiff as wood, and often

I remember exactly the way it looked the day we came up on deck to go ashore: the delicious sparkling tropic sea, the long beautiful beach, the minute palms of the copra plantation waving in the sea breeze, the dark green band of jungle, and the dun mass and power of the mountains rising behind it to rocky peaks.
—NOVELIST JAMES JONES, WHO FOUGHT AT GUADALCANAL

Machine gunners in the jungle.

seven feet high. Walk through kunai grass and your arms and legs will be a mess of cuts. Do you get the picture? Does Guadalcanal sound like a great place to fight? Watch out, you can't even see the enemy hiding in the grass or behind those jungle trees.

The First Marine Division landed in August of 1942. Marines are trained to fight on land or sea. The First Marine Division was a proud division—specially trained, and tough. They needed that toughness. Guadalcanal was one of the hardest-fought battles in history. Remember the back-and-forth slugfest at Gettysburg? This one was worse. It went on for six months. It combined jungle fighting with terrible sea and air battles.

The fight for Henderson Field, the main objective of the entire battle of Guadalcanal. By the end, Japan had lost 600 aircraft and 24 warships. The Americans did not lose much less—but theirs could be easily replaced. Those of the Japanese could not.

At first, things seemed easy. The marines surprised the Japanese on the island, who were mostly construction crews building an airfield. The marines captured the airfield. They named it Henderson Field, after a pilot who had been killed at Midway Island. At last, said President Roosevelt, we have "a toehold in the Pacific."

Words made up from initial letters, like SNAFU, are called **acronyms**. Some other acronyms are WASP ("white Anglo-Saxon Protestant") and NOW (National Organization for Women). Do you know other acronyms?

First Lt. Thomas J. "Stumpy" Stanley became a company commander in the 1st Marine Division. "We thought highly of Stumpy and respected him greatly," wrote E. B. Sledge in a book about the Pacific war called *With the Old Breed*. Tom Stanley was my brother-in-law (and a real hero).

The Japanese were determined to knock that toe into the sea. We wanted to plant both feet on the island. To tell the story of what happened next would take a whole book. Here is some of it:

Let's begin with the military word for a mistake. It is SNAFU, a combination of letters for *Situation Normal, All Fouled Up*. It means that someone goofed.

Who goofed on Guadalcanal? Both sides. It happens all the time in warfare. The pressure and fear of battle often lead to mistakes. Most of our soldiers and even our officers were amateurs. They had not fought in a war before. They had to learn on the job.

One captain, unloading marines onto the beach, didn't want to risk a Japanese attack on his ships. So he pulled out before the loading was finished, taking supplies and marines with him. He stranded the marines already on the island. That was just one of the goofs.

The Japanese officers matched our snafus. They were too sure of themselves. An old Chinese proverb says, "A lion uses all his strength to fight a rabbit." The Japanese were lions in 1942, but they must not have heard of that proverb. They kept sending small forces to Guadalcanal. They thought Japanese fighters were unbeatable. They thought Americans were not good fighters. They were wrong.

When the marines wiped out the first group of Japanese soldiers, their leader was so ashamed he committed suicide. The next Japanese commander arrived on the island with a starched white uniform in his trunk. It was for the surrender ceremony he expected to conduct. After he and his men were destroyed, a marine found the trunk and dressed up in his uniform.

In six months, the marines, and the army units that came to fight with them, lost 1,598 men on Guadalcanal. Japanese war records show an incredible 23,800 deaths. Many Japanese deaths came in suicidal charges. Surrender was considered shameful. The Japanese also suffered many deaths from tropical disease. Our medical care was much better.

Most of the battle for Guadalcanal was fought at sea. There the statistics were more even. Each side lost 24 big ships and many smaller ones. The water near Guadalcanal was so full of sunken ships that it was called "Ironbottom Sound." It should have been called Graveyard Sound. About 20,000 American and Japanese sailors went down there with their ships.

The first of the sea fights—off nearby Savo Island—was the worst disaster in United States naval history. We were whipped. After that it was a seesaw of a conflict. It was bizarre: control switched every 12 hours. The Japanese were skilled fighters in the dark; at night they were mas-

The battles at Guadalcanal, at Bougainville, Tarawa, and in Papua and New Guinea turned marines into seasoned jungle fighters. These Raiders posed in front of a Japanese dugout on Bougainville, near Guadalcanal in the Solomon Islands.

ters of water and air. During the night Japanese planes dropped bombs on Henderson Field; Japanese ships lobbed shells at the field.

In the mornings the Americans took over. Seabees (construction crews) repaired the holes in the airfield. Then our planes took off after Japanese ships and troops. Our pilots were superb during the daytime.

The battles on the island were ferocious. The Japanese were a brave foe. They had not lost a war in 400 years. But the marines outfought them. When it was all over, a doctor examining the marines wrote:

Of 36,000 Japanese soldiers on Guadalcanal when the battle began, 12,000 got home. These men stayed forever.

> *The weight loss averaged about 20 pounds per man....Many of these patients reported being buried in foxholes, blown out of trees, blown through the air, or knocked out.*

The important thing was that they didn't give up. They held Guadalcanal. The marines ended the myth that the Japanese were invincible.

Guadalcanal was a turning point in the war. We went from defense to offense. The Japanese went from offense to defense. A captured Japanese document said, "Guadalcanal Island...is the fork in the road which leads to victory for them or us." We took the right road. It would lead to Japan.

35 Axing the Axis

Rosie the Riveter became the war's can-do symbol, and women took over many jobs usually done by men.

Defense stamps and war bonds were a way to lend the government money. You bought a stamp and stuck it in a book—a bookful of stamps added up to a bond—and cashed the bond in later. Meanwhile, the government used the money you'd paid.

"We have reached the end of the beginning," said Churchill, early in 1943. He was right.

The beginning was horrible. The Germans had perfected their *blitzkrieg* (that lightning attack with planes, tanks, and armies all charging together). The Japanese used the same tactic, and mowed down everyone in their way. The Allies were losing the war. The Axis seemed invincible. Then things began to change. This is what happened:

• In February 1943, Japan pulled out of Guadalcanal.

• That same month, we cracked the Nazi naval code, *Triton*. Now we knew where their submarines were. The Atlantic was full of U-boats, but we began sinking them. The German U-boat admiral couldn't fig-

Below, the Red Army (the Russians) pushed the Germans back from Moscow through the snows of the winter of 1941–1942. A year later, the Germans were trapped in the ruins of Stalingrad (right) and in full retreat.

American women threw themselves into the war effort. Left, polishing airplane noses at Willow Run; above, women welders in a California shipyard. Women also did jobs usually held by men in peacetime: delivering mail, driving buses, and so on.

ure out what was happening.

• We realized that the Germans must have broken our naval code. That would explain why they always seemed to know where our convoys were going. We changed our code. More ships made it to Europe. We began winning the war of the Atlantic.

• The Russians trapped a German army at Stalingrad. Then they laid siege to that army. They starved them. Finally, the German army surrendered. Then the Russians went on the offensive. They headed for Germany. Hitler hadn't planned on that.

• America's factories reached high gear. We began turning out guns, ships, tanks, planes, and other military equipment at an incredible rate—faster than anyone had believed possible. Picture this: a flat, sandy, empty field at a place called Willow Run (in Michigan). Now picture the same field, six months after Pearl Harbor. What you see is a vast building, half a mile long and a quarter of a mile wide. Someone described it as the "most enormous room in the history of man." Steel, rubber, and other raw materials are fed into one end of the room; airplanes emerge from the other end—almost 9,000 airplanes the first year. It is not surprising that many historians say the Second World War was won in America's factories and laboratories.

• In 1943, the Russians were fighting the Axis alone on the European continent. Stalin was crying for help. He asked his allies to land forces in Europe and take some pressure off his troops. He asked for a second front. American and British leaders agreed,

A **front** is a battle line. Stalin's front was in Russia and eastern Europe. He wanted the Allies to launch a western front, so that the Germans would have to fight on two sides at once.

Convoys were groups of ships. Ships carrying troops or supplies traveled together with destroyers for protection against submarines.

Chips, who is part husky and part German shepherd, is awarded the Army's Distinguished Service Cross for "courageous action in single-handedly eliminating a dangerous machine-gun nest" and for capturing four Italian gunners in Sicily. Later the award is withdrawn when the War Department rules that dogs are not eligible for medals.

"The venture [in North Africa] was new," said General Eisenhower (above, right, with General Patton). "Up to that moment no government had ever attempted to carry out an overseas expedition involving a journey of thousands of miles from its bases and terminating in a major attack." Right, U.S. troops in North Africa.

In 1942, when Rommel (on the right, in Tripoli) reported to Hitler and Hermann Göring that Britain was dropping American shells on his men, Göring said, "Quite impossible. All the Americans can make are razor blades and refrigerators." Rommel replied, "I wish, *Herr Reichsmarschall*, that we had similar razor blades!"

and made plans for a joint landing. Its code name was *Operation Torch*.

But when the landing came, it was in North Africa, not Europe. That wasn't exactly what Stalin wanted, but it did help. North Africa was a good place to begin our offensive. It had been 23 years since we fought in World War I, and our troops needed combat experience. North Africa became a war school for us, with General Dwight D. Eisenhower in charge.

The Nazi forces there were led by General Erwin Rommel, who was known as the "desert fox." Rommel was intelligent, wily, and tough. His Afrika Korps had been destroying British troops. Then the British went on the offensive, heading west from Egypt. (Massive supplies helped.) Combined Allied forces headed east from Morocco and Algiers. A small French force came north from Chad. Rommel was caught in a pincer. We had managed to outfox the fox. We drove the Axis from North Africa. The Germans no longer controlled the Mediterranean Sea.

• We were now bombing Germany from the air day and night, but we needed to do more than that. We had to invade and help destroy Hitler's forces. Should we land in France and push east to Germany? Should we land in Italy and move north? Should we go through Greece? Finally, it was decided. We would start on the Mediterranean island of Sicily and go on to Italy. Look at a map and you'll see why Sicily was important.

• The invasion of Sicily was given the code name *Husky*. We landed by sea and air and captured the island. Amphibious ducks were used for the first time. But there was a snafu: we let an Axis army escape to Italy.

• The Italian people were now fed up with war. Our bombs were blasting Rome. Things hadn't worked out as some Italians thought they would. Their morale collapsed. They kicked Mussolini out of power. Their army went home. They got out of the war. Because of that, we thought Italy would be easy to capture. It wasn't. More snafus. The German army moved into the mountainous Italian peninsula and captured the mountaintops. Picture the enemy shooting down at you as you attempt to climb. That's what happened. The Germans were on the heights. Our soldiers faced ferocious fire. Italy was a bloody standoff.

• Planning began for *Operation Overlord*—code name for the invasion of France. It was to be the largest amphibious invasion in all of history. The Nazis knew it was coming, so they began planning to defeat it. We assembled men and materials in England. The Germans readied their defenses. They laid explosive mines all along the coast, layers and layers of mines. Then they put steel and concrete barriers in the water and on the beaches. They added barbed wire, huge steel spikes, and more mines on the beaches. They built rooms of thick concrete—called *bunkers*—and filled them with heavy antitank guns, versatile medium-size guns, deadly flamethrowers, and machine guns. They called all this the *Atlantic Wall*.

Top, U.S. troops pass beneath the Colosseum as they enter Rome. Center, Romans give the liberating Americans a cheerful welcome. Bottom, Rommel's ferocious Atlantic Wall on the beach at Pas de Calais.

They fortified the whole coastline—from the Netherlands to the west coast of France—although they were sure they knew the exact spot where the landing would be made. Everyone knew. Look at the map on page 160 and see if you can figure it out too.

The best route is obvious: from Dover, England, to the Pas de Calais (pah-duh-KAL-ay) in France. It is the shortest distance across the treacherous English Channel, it has gentle beaches, and it is the best place to land if you are heading for Germany's heartland. German intelligence officers decoded Allied messages that told of invasion plans for Calais. German pilots bombing England brought back photographs of tanks and trucks lined up near Dover. Spies reported on plans for a second invasion, on the same day, into Norway.

The messages were all fakes. The tanks and trucks were big balloons, designed to look real from aerial photographs. We weren't going to land at Pas de Calais. We weren't planning a Norway invasion. The German spies were double agents working secretly for us.

155

The End of the Road

Ernie Pyle was a war correspondent. In this excerpt from his book Brave Men, *he is writing about the war in Italy.*

Ernie Pyle (center, balding) on Iwo Jima early in 1945. He died the same year, shot on the island of Ie Shima.

Captain Waskow was a company commander....He was very young, only in his middle twenties, but he carried in him a sincerity and gentleness that made people want to be guided by him.

"After my father, he came next," a sergeant told me.

"He always looked after us," a soldier said. "He'd go to bat for us every time."

"I've never known him to do anything unfair," another said.

I was at the foot of the mule trail the night they brought Captain Waskow down. The moon was nearly full and you could see far up the trail....Dead men had been coming down the mountain all evening, lashed onto the backs of mules. They came lying belly-down across the wooden pack-saddles, their heads hanging down on one side, their stiffened legs sticking out awkwardly from the other, bobbing up and down as the mules walked.

I don't know who the first one was. You feel small in the presence of dead men, and you don't ask silly questions....We left him there beside the road, that first one, and we all went back into the cowshed and sat on water cans or lay on the straw, waiting for the next batch of mules....We talked soldier talk for an hour or more....Then a soldier came into the cowshed and said there were some more bodies outside. We went out into the road. Four mules stood there in the moonlight, in the road where the trail came down off the mountain. The soldiers who led them stood there waiting.

"This one is Captain Waskow," one of them said quietly.

Two men unlashed his body from the mule and lifted it off and laid it in the shadow beside the stone wall. Other men took the other bodies off. Finally, there were five lying end to end in a long row....The unburdened mules moved off to their olive grove. The men in the road seemed reluctant to leave. They stood around, and gradually I could sense them moving, one by one, close to Captain Waskow's body. Not so much to look, I think, as to say something in finality to him and to themselves. I stood close by and I could hear.

One soldier came and looked down, and he said out loud, "God damn it!" That's all he said, and then he walked away.....

Another man came. I think he was an officer. It was hard to tell officers from men in the dim light, for everybody was bearded and grimy. The man looked down into the dead captain's face and then spoke directly to him, as though he were alive, "I'm sorry, old man."

Then a soldier came and stood beside the officer and bent over, and he too spoke to his dead captain, not in a whisper but awfully tenderly, and he said, "I sure am sorry, sir."

Then the first man squatted down, and he reached down and took the captain's hand, and he sat there for a full five minutes holding the dead hand in his own and looking intently into the dead face. And he never uttered a sound all the time he sat there.

Finally he put the hand down. He reached over and gently straightened the points of the captain's shirt collar, and then he sort of rearranged the tattered edges of the uniform around the wound, and then he got up and walked away down the road in the moonlight, all alone.

The rest of us went back into the cowshed, leaving the five dead men lying in a line end to end in the shadow of the low stone wall. We lay down on the straw in the cowshed, and pretty soon we were all asleep.

36 Going for D-Day

June 6, 1944, England: Eisenhower gives paratroopers leaving for Normandy the orders for D-Day—to go all-out for a full victory.

General Erwin Rommel, Germany's brilliant desert warrior, looked at the skies and decided he would take a two-day trip home to Germany. It was his wife's 50th birthday (he had bought her a gift—shoes from Paris); he also wanted to see the Führer and ask for more troops. The weather was too rough for an invasion, he said.

General Eisenhower looked at the same skies and decided to go for it. The English Channel was in turmoil, but the moon was full and the tides were low.

The invasion began at night, when paratroopers dropped behind enemy lines. They captured bridges and lit flares to guide the gliders that followed. With all the sophisticated equipment available, it was a tiny child's toy—a snapper that made a sound like a cricket—that the paratroopers used as a signal during the night so they could find each other.

Then, at daybreak, the sky filled with airplanes—wingtip to wingtip—9,000 of them. Two submarines raised flags to mark a landing area. The largest armada ever assembled appeared off the French coast: landing vehicles, minesweepers, attack transports, tankers, cruisers, battleships, ocean liners, yachts, hospital ships and puffing tugs—all the boats and ships that could be

Hitler was known in Germany as *Der Führer* (FEW-rur), which means "the leader" in German.

Troops plunge down their Coast Guard landing barge ramp and wade for the beach. Behind the clouds loom the Normandy cliffs they will climb.

General Rommel had good reasons for taking time off. He intended to see Hitler and ask for more men to beef up the beach defenses. He also wanted to see Manfred, his 14-year-old son, who had been drafted into the army. Some of the other officers who took time off got together to practice war games.

157

Antiaircraft barrage balloons over Omaha Beach. Among the arrivals on June 7, D-Day-plus-1, are a group of hulks that will be sunk off the beach as the foundations for temporary artificial harbors. The Allies need a base from which their invading troops can be easily landed and their equipment and supplies handled.

found. They made an awesome fleet 20 miles wide. Giant military barrage balloons floated above, to interfere with enemy planes.

It was June 6, 1944, and forever it would be known as *D-Day*. The Allies were heading for the treacherous, mine-strewn beaches of Normandy in France, 100 miles from the nearest English port across the turbulent Channel. Enemy soldiers, in bunkers on top of the Normandy cliffs, some of them 150 feet high, waited behind formidable heavy guns.

But most were asleep. No one was expecting an attack in this weather. Rommel wasn't the only officer on

In England, General Eisenhower's weather-man—tall, serious-faced Captain James M. Stagg —had reports from five weather stations in the Atlantic. The reports indicated that there would be a 16-hour "window" in the bad weather. Ike entered Europe through that window.

Soldiers are supplied with assault equipment suited to their landing area. Those on the cliffs have ropes and ladders; these men have bikes.

At Pointe du Hoc, which rises 100 sheer feet from the beach, U.S. Rangers climb up the cliffs on wire ladders as bullets and grenades rain down on them. Directly above is a German gun battery.

A Great War Photographer

Most of Robert Capa's Normandy photographs were never published; they were accidentally ruined by a Life magazine darkroom technician.

vacation. Most of the German leaders had taken the weekend off.

What happened next? The landing had been planned with the precision of a ballet. Everyone had a place and time in the drama. And, at four of the five landing beaches, things went more or less on schedule. But on Omaha Beach (one of two beaches where Americans landed), everything seemed snafued. The first men ashore couldn't secure the beach. What was supposed to take minutes took hours. Of 32 tanks, with collars that were supposed to keep them afloat, 27 sank in the choppy water with men inside. Allied planes, sent to bomb the enemy's gun-filled bunkers, went too far, missed the guns, and dropped their bombs on French cows. Immense traffic jams of men and supplies backed up in the water and on the beaches. Mines and shells were exploding everywhere. Gliders dropped men and supplies behind the beaches into swampland, where many sank. "Our men simply could not get past the beach. They were pinned down right on the water's edge by an inhuman wall of fire....Our first waves were on the beach for hours, instead of a few minutes, before they could begin working inland," wrote war correspondent Ernie Pyle, who was there.

General Omar Bradley, on the command ship *Augusta*, thought about calling off the landing. Then a destroyer came up into the shallow water and lobbed a shell right inside a main bunker. When other ships added their firepower, the Nazi gunners in the concrete

Robert Capa was one of the war's great photographers. He took part in the D-Day invasion: *The sea was rough and we were wet before our barge pushed away from the mother ship. It was already clear that General Eisenhower would not lead his people across the Channel with dry feet or dry anything else. In no time, the men started to puke. But this was a polite as well as a carefully prepared invasion, and little paper bags had been provided for the purpose. [There weren't enough of the bags; most men used their helmets.]*

General Omar Bradley (right), U.S. 1st Army commander, and British General Bernard Montgomery discussing maneuvers in a field near Omaha Beach.

Emplacements are the areas where heavy artillery (guns) are positioned.

159

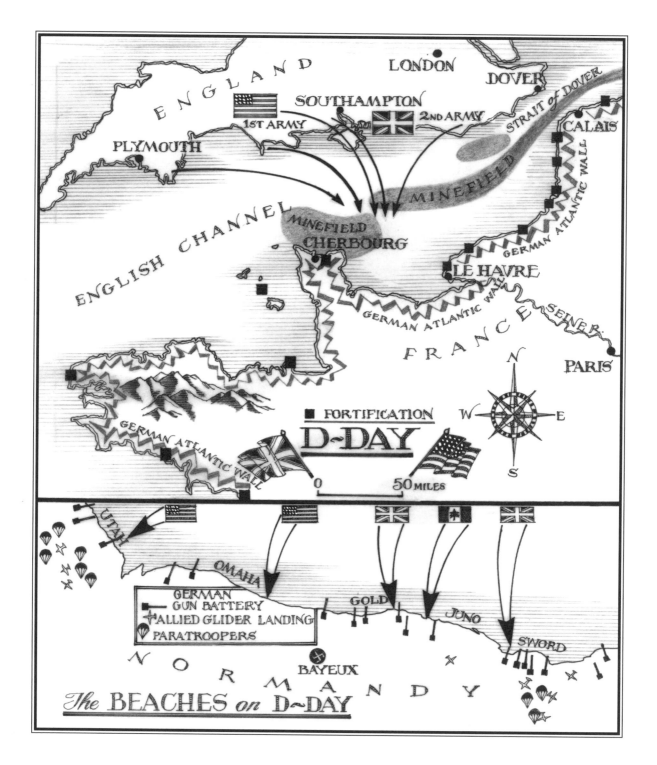

The BEACHES on D~DAY

emplacements were in big trouble. The Navy had opened a crack in the German defenses and the Yanks were on their way.

Slowly at first, but then steadily, soldiers and medical personnel began to land and head inland, into the fierce guns on top of the bluffs. No word is big enough to describe their effort. *Heroic* will have to do.

The operation had been brilliantly planned. Troops and officers had trained for a year; it paid off. Equipment specially designed for this invasion worked superbly. Tanks unrolled reels of steel matting that made roadways across the sand. Other tanks, with flailing chain arms, detonated mines and began to make the beach safe. Some tanks carried small bridges. Naval engineers had built huge floating harbors; they were towed into place.

Normandy's hedgerows (above) were ancient banks of earth topped by bushes. It was hard for tanks to get through, so a U.S. sergeant came up with these steel prongs to attach to the front of the tank, made from discarded German beach obstacles.

In the midst of the fighting, over the noise of battle, a British major shouted out words from Shakespeare's play *Henry V*, about another invasion of France.

> *We few, we happy few, we band of brothers;*
> *For he today that sheds his blood with me*
> *Shall be my brother; be he ne'er so vile,*
> *This day shall gentle his condition.*
> *And gentlemen in England now a-bed,*
> *Shall think themselves accursed they were not here;*
> *And hold their manhoods cheap whiles any speaks*
> *That fought with us upon Saint Crispin's day.*

It was D-Day, not St. Crispin's, but poets would write of this day too: of heroism, of achievement, and of the waste of war. It was there to see on the beaches. They were littered with tanks and bodies and the leftovers of men's lives: socks, Bibles, toothbrushes, diaries, mirrors, letters, first-aid kits, photographs, and food rations. From Ernie Pyle:

> *There was a dog...on the water's edge, near a boat that lay twisted and half sunk at the waterline. He barked appealingly to every soldier who approached, trotted eagerly along with him for a few feet, and then, sensing himself unwanted in all the haste, he would run back to wait in vain for his own people at his own empty boat.*

By nightfall, Allied troops—American, British, Canadian, Free French, Polish—were holding French soil. We had made it. We were on our way to Berlin.

For the Axis, it was the beginning of the end.

With a firm foothold on French soil, U.S. infantrymen march through a Normandy town past a wrecked German truck still in its leafy camouflage.

37 A Wartime Diary

The Battle of the Bulge was Hitler's last gamble on the western front. Much of it was fought in bad, snowy conditions. Above, GIs in a chow line.

Find the Black Sea—above Turkey and below Russia—and you'll find the Crimean peninsula and Yalta.

You are a newspaper writer, a war correspondent, covering this world war. When it ends you will write a book about the times you have lived through, so you have been keeping a diary to record events. It is 1945. Here are some excerpts from your journal, and some notes from this author:

JANUARY 1: Allied forces are beginning to turn back a powerful German army in Belgium. Soldiers are calling this the Battle of the Bulge. *A strong German offensive created a bulge in the Allied defensive line.* Whatever you want to call this battle, it is fierce. Still, it looks as if this will be a good year for the Allies. Germany's armies are being battered on the ground, its cities pounded from the air.

JANUARY 20: FDR is inaugurated for a fourth term as president. No other president has served more than twice.

JANUARY 24: Russian soldiers have crossed the Oder River and, for the first time, are on German soil.

JANUARY 27: Today, in Poland, Soviet forces liberated a German concentration camp. It is named Auschwitz. People in the camp were used as slaves by a German industrial chemical company. The company has left charts of the costs and profits of slave labor. But the camp was built primarily for another purpose. It was built to kill people. The horror and evil of it all are more than anyone

Christmas Eve, 1944: an 82nd Airborne Division infantryman on a one-man sortie through the lines during the Battle of the Bulge. The bump on the horizon is the head of the sniper covering him.

Left: the Big Three at Yalta. But their unity did not go far beyond beating Hitler, as this cartoon, showing Stalin with "secret plans," suggested.

wants to believe. *Three million people were murdered at Auschwitz, most of them Jews.*

FEBRUARY 1: The sky is thick with planes—an incredible sight! One thousand airplanes are flying over Europe on their way to bomb Berlin.

FEBRUARY 4–11: President Roosevelt has flown to Yalta, in the Crimea, to meet with England's Winston Churchill and Russia's Joseph Stalin. They are said to be planning for the final battles of the war. They agree to call a meeting of the world's nations in San Francisco in April, to form a peace-keeping organization called the United Nations.

FEBRUARY 14: John D. Rockefeller, Jr., donates $8.5 million to buy land in New York City as a permanent home for the United Nations. Woodrow Wilson's idea of a league of nations may finally be achieved.

ABOVE, two days before the German army surrenders, U.S. troops force German citizens to inspect the bodies of victims of a Nazi concentration camp near their town.

January 1945: Red Army soldiers, now only 50 miles from Berlin, struggle to get a field gun across the river Oder.

Below, at Luzón, MacArthur returns to the Philippines. He waded (instead of rowing in) on purpose, to look grander. And he made the photographer do the shot three times before he thought it was perfect.

FEBRUARY 19: American marines have landed on Iwo Jima, an island in the Pacific. Reports tell of incredibly hard fighting and great losses on both sides.

FEBRUARY 24: General Douglas MacArthur promised to return to the Philippines—and he has done it.

MARCH 7: The U.S. 1st Army has crossed the river Rhine and is inside Germany. All German armies have been pushed back into the Fatherland.

MARCH 9: Today, 325 low-flying bombers dropped incendiary bombs on Tokyo. The bombs are filled with jellied gasoline meant to set fires. The Japanese capital became so hot that the water in its canals boiled. Still the Japanese refuse to surrender. *Later reports list 267,000 Tokyo buildings burned to the ground; 89,000 persons dead.*

MARCH 21: Today we watched an enormous air armada fly overhead on its way to Germany. *There were 7,000 Allied planes.* Until now, most bombing raids have been at night, to make it difficult for the antiaircraft guns and fighter planes to see the bombers. Now Germany has little firepower left; the bombers fly at will. Bomb damage there is said to be devastating.

The Axis powers now have no hope of winning and yet neither Germany nor Japan will give up. Those who began this insane engine of war have cut out the brakes. They can't seem to stop. It is their own people who are suffering most. Do the warlords care?

Left, the war's most famous photograph: raising the flag on Mt. Suribachi after the capture of the island of Iwo Jima. Below, U.S. firebombs fall on Kobe, Japan.

38 April in Georgia

Much wartime propaganda, like this American postcard, was meant to reassure the troops, and the folks at home, that their efforts were succeeding and that the war was worth fighting.

The madmen who run Japan and Germany refuse to give up. They talk of leading their nations in a fight to the death. Terrible firebombs are dropping on Tokyo and Dresden. Hundreds of thousands of people are dying. Just as a fist squeezes its contents, so British, American, and Russian troops are squeezing Germany. In the Pacific theater we are making plans to invade Japan. Everyone expects that invasion to be bloodier than the one in Normandy.

The Allies will win this war—that now seems clear—but the German and Japanese leaders are making it very difficult. Like ancient rulers who had their followers killed and buried in their tombs, these leaders seem determined to kill their own people.

The president knows of something that might end the war quickly. It is that secret weapon that almost no one else knows about. Partly because of this, he feels he can relax and catch up on some paperwork. He is exhausted. He has just turned 63, but he looks much older. The war has been a terrible strain: he has traveled around the world, he has run for a fourth term as president, he has

Where is Tokyo? Where is Dresden?

Contemplating the results of a firestorm set off in Dresden, Germany, by Allied bombing in February 1945.

165

The exhaustion brought on by the war years shows plainly in FDR's face as he arrives in the Crimea for the Yalta conference. He was a sick man.

On the island of Okinawa, the U.S. Army suffers some 80,000 casualties. Japanese losses total 120,000.

Roosevelt is planning to attend a Jefferson's birthday celebration. Do you remember another famous party for that occasion? It was attended by Andrew Jackson and John Calhoun. (See Book 6 of *A History of US* for details.)

been active as commander in chief of the armed services, he has been an inspiring leader. He needs to take it easy for a few days. He makes plans to go to Warm Springs, Georgia. He first visited Warm Springs years earlier, when he was recovering from polio. The waters are healing. He has been back many times, and has grown to love the slow-paced gentleness of the South.

In Georgia, wild violets are blooming; so are purple-blue wisteria and sweet-smelling honeysuckle. It is springtime—April 12, 1945—and, at Warm Springs, cooks are preparing a picnic. The smells of barbecued beef and chicken fill the air. As the president works, an artist sits nearby, making sketches for a watercolor portrait.

Almost exactly 80 years earlier, another American president had decided to relax and go to the theater, knowing that a terrible war was coming to an end.

Like that other president, Roosevelt is concerned about the peace that is to come. He wants this war to have meaning. Soon after the war began, he met with Winston Churchill and signed a document called the *Atlantic Charter*. It says that after the war, nations will be free to choose their own forms of government. That is

The president on the porch of his home in Warm Springs, known as the Little White House. It is still there.

called *self-determination*. Roosevelt wants to end the old, before-the-war imperialist ways. Then, a few European nations ruled much of the world. Sometimes they ruled well; sometimes not well. To Roosevelt, that doesn't matter now. People should be free to govern themselves. Great Britain still controls India and Burma. France expects to regain control of Indochina (which includes Vietnam). Japan has attempted to become an empire. The United States rules the Philippine Islands.

Roosevelt thinks imperialism—even well-meaning imperialism—is wrong. He will show the world: America has no desire for other lands. We will begin by granting independence to the Philippines. He is planning to go to the independence ceremonies himself.

Russia is a worry. The Russian people have fought magnificently. They have been brave allies. But they aren't a free people. Stalin is a dictator. Winston Churchill believes that Stalin cannot be trusted.

Roosevelt is beginning to have worries about Stalin too.

At Warm Springs he works on a speech to be given at the dinner to honor Jefferson's memory. This is part of what Roosevelt writes:

> The once powerful, malignant Nazi state is crumbling. The Japanese war lords are receiving, in their own homeland, the retribution for which they asked when they attacked Pearl Harbor.
>
> But the mere conquest of our enemies is not enough.
>
> We must go on to do all in our power to conquer the doubts and the fears, the ignorance and the greed, which made this horror possible....if civilization is to survive, we must cultivate the science of human relationships—the ability of all peoples, of all kinds, to live together and work together, in the same world, at peace....
>
> The work, my friends, is peace. More than an end of this war—an end to the beginnings of all wars. Yes, an end, forever, to this impractical, unrealistic settlement of the differences between governments by the mass killing of peoples.

The president is sitting in a leather armchair; he turns to the artist. "We've got just 15 minutes more," he says. Some cousins of his and a friend are in the room. They are quiet. The president is studying papers. The 15 minutes are almost up when he raises a hand to his temple. "I have a terrific headache," he says. They are the last words he will ever speak.

The August 1941 meeting between Roosevelt and Churchill aboard the U.S.S. *Augusta* that paved the way for America's entry into World War II.

Malignant means harmful.
Retribution is punishment.

Our only hope will lie in the frail web of understanding of one person for the pain of another.
—JOHN DOS PASSOS, DECEMBER 1940

What do you think he means by that? Do you agree?

The last family photo taken before Roosevelt's death: FDR, Eleanor, and their 13 grandchildren at the White House on Inauguration Day, 1945.

167

39 President HST

In 1918 Captain Harry Truman commanded a field-artillery battery in France, where he saw active combat.

The solidly built, gray-haired man with metal-rimmed glasses sat at a high desk facing the Senate floor. A gold-bordered blue velvet drape hung behind him and made a frame for his chair. Vice President Harry S. Truman was presiding over the Senate. At least that was what he seemed to be doing. Actually, he was bored and was writing a letter to his mother and sister, who were back in his hometown of Independence, Missouri. This is part of what he wrote:

Dear Mama & Mary: I am trying to write you a letter today from the desk of the President of the Senate while a windy Senator...is making a speech on a subject with which he is in no way familiar....Turn on your radio tomorrow night at 9:30 your time....I think I'll be on all the networks....I'll be followed by the President, whom I'll introduce.

It had amazed almost everyone—including Truman—when he was asked to be vice president. He hardly knew FDR. Some said that it was the Democratic National Committee that selected Harry Truman; that the other

Truman was asked to be vice president by Democratic National Chairman Robert Hannegan, but he refused—said he wasn't a candidate. Just then the phone rang. It was FDR. He asked Hannegan if he had "got that fellow lined up yet?"

"He is the contrariest Missouri mule I've ever dealt with," said Hannegan.

"Well, you tell him," FDR shouted, and Truman could hear him, "that if he wants to break up the Democratic Party in the middle of a war, that's his responsibility."

Truman with friends in his Kansas City, Missouri, men's clothing and haberdashery store, around 1920.

Election night in 1944: FDR (left), returning for a fourth term as president, with his outgoing vice president, Henry Wallace (right), who seems to be enjoying a joke with the man taking over his job, Harry Truman. Truman was in office as vice president for only four months.

THE SPOKESMAN-REVIEW

ROOSEVELT IS DEAD, TRUMAN SWORN IN

President Dies Without Pain; Had Complained of Bad Headache

candidates considered were all controversial.

Truman didn't seem to have any enemies. But he didn't have many enthusiasts either. He had become a senator at age 50. He'd fought in the Great War, and had tried a number of jobs—he'd even owned a men's clothing store—but the store had failed, and he wasn't very successful at anything until he got into politics, first as a postmaster and then as a judge. He was a bookish sort, with an astounding knowledge of history: quiet, honest, likable, and fair-minded, although he sometimes lost his temper. It was as head of a Senate committee investigating military contracts that he had impressed people—including Roosevelt. His committee probably saved the government billions of dollars.

When the windy senator finally finished, Truman went to visit his old friend, the speaker of the House of Representatives, Texan Sam Rayburn. Rayburn's office was a gathering place where congressmen relaxed and gossiped. It was called—in jest—the Board of Education. It was there that Truman got a call from the president's press secretary, Steve Early. He was to come to the White House at once. Early's voice had an urgent tone.

Truman headed through the underground passage to the Senate Office building—there his Secret Service agents lost track of him. But his car and driver were waiting. As he drove the 15 long blocks to the White House, he guessed that the president had flown in from Georgia and wanted him for something ceremonial.

Upstairs at the White House, he learned differently. Eleanor Roosevelt put her hand on his shoulder and said softly, "Harry, the president is dead." For a moment he could say nothing. Then he asked if there was anything he could do for her.

"Is there anything we can do for you?" she answered. "You are the one in trouble now."

Truman's wife, Bess, looks on as her husband, in his trademark round glasses and bow tie, is sworn in as the 33rd president of the United States.

169

40 A Final Journey

In his fireside chats, [Roosevelt] talked like a father discussing public affairs with his family in the living room.
—WILLIAM E. LEUCHTENBERG

"Mr. Roosevelt was great," a Harvard professor said to his students, "because he, like Lincoln, restored men's faith."

The poem Eleanor Roosevelt remembered was written by Millard Lampell.

It was as if a member of the family had died. He had been a world-dominating figure for 12 years—strong, witty, compassionate, able. Young people could remember no other president. The nation was in a state of shock.

Those who had felt left out by government before—the poor and disadvantaged—were especially grieved. This was a president who had done more than talk about fairness and opportunity; he had acted to begin to make them reality. And, while he had broken precedent by running for office four times, he had not forgotten that as president he was the servant of the people. He had never assumed kingly trappings. He had never lost his sense of humor or his easy informality.

In Warm Springs, the flag-draped coffin began its long, sad train journey—to Washington first, and then to Hyde Park, where the president was to be buried. Sitting inside the train, Eleanor Roosevelt kept remembering a poem about Lincoln's death. It wouldn't leave her mind:

> *A lonesome train on a lonesome track*
> *Seven coaches painted black,*
> *A slow train, a quiet train*
> *Carrying Lincoln home again.*

On April 13, 1945, a hearse rolled into Warm Springs train station. Franklin Delano Roosevelt, 32nd president of the United States, was going home to Hyde Park for the last time.

At night, unable to sleep, she said:

I lay in my berth…with the window shade up, looking out at the countryside he had loved and watching the faces of the people at stations, and even at the crossroads, who had come to pay their last tribute all through the night….I was truly surprised by the people along the way; not only at the stops but at every crossing.

In the train's press car, reporters looking out the window saw black sharecroppers on their knees, hands outstretched in prayer. As the train slowed in a South Carolina city, members of a Boy Scout troop began singing "Onward, Christian Soldiers"; then others joined in, and soon, according to one who was there, "eight or ten thousand voices were singing like an organ." Everywhere people cried. The sobs continued as the coffin was carried, by horse-drawn caisson, through Washington to the White House. Then, during the memorial service, the whole grieving nation came to a halt and paid its respects.

Airplanes sat on runways; radios were silent; telephone service was cut off—there was not even a dial tone; news-service teletypes typed the word *silence* and went dead; movie theaters closed; cars and buses pulled to the curb; 505 New York subway trains stopped; stores shut their doors; and everywhere—in other countries too—people put hands on hearts, or fell to their knees, or just stayed quiet. That day newspapers carried no advertisements.

Clearly he had been a great president. But how great? What would history say of him?

A poll of 50 leading historians soon ranked him just behind Lincoln and Washington as one of the three greatest presidents. Winston Churchill said that in world importance, Roosevelt was first.

And yet, as much as some loved and respected him, others hated and vilified him.

Later historians would look at him through two lenses. As an effective president, they agreed, he was like a magnificent symphony conductor who knows all the notes and just what to do with them. No question about it, they still agreed, he was a great president. But there was something that bothered many. As a human being, he was sometimes less than great. His personality was flawed. It was too bad to have to acknowledge it, but he could be devious. He could tell a person something and not mean it. He could tell a story that made him look good, but wasn't quite true.

Perhaps it was that tendency to always act as if everything was fine, even when it wasn't, that some found disturbing. He was used to pretending. It was both a strength and a weakness.

His wife said:

"I felt as if I knew him," said one young man. "I felt as if he knew me—and I felt as if he liked me."

After the president's death, Eleanor Roosevelt said that people often told her how "they missed the way the president used to talk to them…[in his radio fireside chats]. There was a real dialogue between Franklin and the people," she reflected. "That dialogue seems to have disappeared from the government since he died."

He was in a very special sense the people's president, because he made them feel that with him in the White House they shared the presidency. The sense of sharing the presidency gave even the most humble citizen a lively sense of belonging.

—JUSTICE WILLIAM O. DOUGLAS

171

Faces along the route of the president's funeral cortège.

Because he disliked being disagreeable, he made an effort to give each person who came in contact with him the feeling that he understood what his particular interest was....Often people have told me that they were misled by Franklin....This misunderstanding not only arose from his dislike of being disagreeable, but from the interest that he always had in somebody else's point of view and his willingness to listen to it.

He listened intently, and that was flattering. People thought it meant that he agreed with them. He didn't tell them differently.

So some felt he couldn't be trusted.

But no one could take his achievements as president from him. They changed the nation. Here are the most important of them:

• He was a peerless crisis manager. He led the nation through two of its worst times—a depression and a world war—with gusto, courage, and an unfailing confidence.

The only thing we have to fear is fear itself.

• He was inspiring. He made people believe in their country and want to do their best for it. Perhaps only during that time when the Constitution was written were more brilliant thinkers attracted to government service.

This generation has a rendezvous with destiny.

• He believed in government for the people. During the Roosevelt administrations, Social Security, farm programs, aid for home buyers, aid for dependent children, and other caring programs were begun. He paid attention to laboring people and their needs. Some called it a "revolution." Perhaps it was. It was in line with a tradition of revolution that could be traced to the ideas of Thomas Jefferson, Andrew Jackson, the Populists, and the Progressives.

As we have recaptured and rekindled our pioneering spirit, we have insisted that it shall always be a spirit of justice, a spirit of teamwork, a spirit of sacrifice, and, above all, a spirit of neighborliness.

• He strengthened the two-party system.

The war casualty announcement, which listed daily those who had died in the military services, was headed on April 13 by the name of Franklin D. Roosevelt.

Since before the Civil War, only two Democrats —Grover Cleveland and Woodrow Wilson—had been elected president. (And Woodrow Wilson made it because the Republican Party was split.) After Roosevelt there was a better balance between the parties.

Here in the United States we have been a long time at the business of self-government. The longer we are at it the more certain we become that we can continue to govern ourselves; that progress is on the side of majority rule; that if mistakes are to be made we prefer to make them ourselves and to do our own correcting.

He brought new people into government. He named a woman, Frances Perkins, to his cabinet. He began the process to "include the excluded." (Wartime America was a land that accepted much discrimination.)

We are going to make a country where no one is left out.

• He cared about the environment. He sent young people from the inner cities out to plant trees, and he worked to protect our nation's natural heritage.

The conservation of our natural resources and their proper use constitute the fundamental problem which underlies almost every other problem of our national life….the government has been endeavoring to get our people to look ahead and to substitute a planned and orderly development of our resources in place of a haphazard striving for immediate profit….We are prone to think of the resources of this country as inexhaustible; this is not so.

• By his personal example he showed that—for people with energy and intelligence—there need not be any such thing as a handicap.

If you have spent two years in bed trying to wiggle your big toe, everything else seems easy.

• He won the war and set the stage for the prosperity that was to follow. It might not have happened with another leader.

In the future days, which we seek to make secure, we look forward to a world founded upon four essential freedoms.

The first is freedom of speech and expression—everywhere in the world.

The second is freedom of every person to worship God in his own way—everywhere in the world.

The third is freedom from want….The fourth is freedom from fear.

"We have learned," President Roosevelt said at his fourth inauguration, "to be citizens of the world, members of the human community. We have learned the simple truth, as Emerson said, that 'the only way to have a friend is to be one.'"

41 Day by Day

April 29: U.S. troops reach Dachau concentration camp in southern Germany. An inmate writes, "First American comes through the entrance. Dachau free!!! Indescribable happiness. Insane howling."

More from the diary that you, the newspaper reporter, are keeping.

APRIL 12, 1945: American soldiers enter the Nazi concentration camp at Buchenwald and find death everywhere. Germans interviewed say they didn't know anything about the concentration camps. *As camp after camp is discovered, the Allies react with cold fury. More than 10 million human beings have died in the concentration camps. In one camp American soldiers discover bins with thousands of pairs of babies' shoes.*

APRIL 13: Everywhere there is shock and disbelief. The president is dead! It is hard to imagine the United States without President Roosevelt. But the bombing of Germany continues—day and night. Germany's cities are in ruins. Mostly the bombs kill civilians. Are they to blame for this war?

This bombing of cities is called strategic bombing. *There is something about it that is strange. The tougher the bombing, the more it makes people want to work hard and fight back. When the German Luftwaffe was bombing England, it seemed to inspire the English to fight and produce. All selfishness was forgotten. Military production went up. The same thing happened in Germany and Japan. After last year's fierce raids on Germany, there was a short lull and then aircraft production greatly*

The Camps

In 1945, Leon Bass was a 19-year-old soldier attached to the American 3rd Army when his unit was sent to Weimar, Germany.

Immediately about five or six of us took off with an officer to a place called Buchenwald....Buchenwald was a concentration camp. I had no idea what kind of camp this was. I thought it might have been a prisoner-of-war camp where they kept soldiers who were captured. But on this day...I was to discover what human suffering was...about. I was going to take off the blinders that caused me to have tunnel vision. I was going to see clearly that, yes, I suffered and I was hurting because I was black in a white society, but I had also begun to understand that suffering is universal. It is not just relegated to me and mine; it touches us all.

Anne Frank

increased. The military experts don't want to believe it, but it seems to be true. The military experts say the purpose of the bombing is to destroy morale.

APRIL 14: The British liberate the Bergen-Belsen concentration camp. *On March 12, in this camp, a 14-year-old Dutch girl died. Her name was Anne Frank. After the war ends her diary will be found and published.*

APRIL 20: Hitler is 56 today. Is there anyone left who wishes him a happy birthday? *Probably. His strongman ideas had wide appeal. Democracy is difficult; it asks people to think and take part in their government. The totalitarian governments treat people like sheep—or is it dogs?*

APRIL 21: Russian troops have entered the suburbs of Berlin, Germany's capital. The fighting is said to be fierce, although the Nazis have no hope

At war's end, many German cities had been bombed to ash and rubble. Here, Germans search for usable bricks to begin to rebuild Dresden.

of victory. *In less than one month, between April 16 and May 8, the Russians lose 304,887 men—killed, wounded, and missing—capturing Berlin. The total number of American deaths in the whole war, in Europe and the Pacific, is about 325,000.*

APRIL 24: If only there were a way to end this war quickly. The deaths now seem so unnecessary. *On this day President Truman gets his first detailed briefing on the top-secret superweapon.*

APRIL 26: Italy's dictator, Benito Mussolini, has been hanged by Italians fighting on the side of the Allies.

APRIL 30: Hitler is dead. He has killed himself. He was living like a mole underground in a concrete bunker in Berlin.

LEFT, top: Mussolini in death. Bottom, the mayor of Leipzig, Germany, and his wife and daughter have committed suicide as U.S. infantry close in on the city. ABOVE, Russian soldiers approach the ruined Reichstag in Berlin. "Wherever we looked we saw desolation," said an American general.

Left, General Alfred Jodl signs the German surrender at Rheims. The war in Europe is over. None of Eisenhower's staff has a great idea for a victory statement, so the general writes his own: "The mission of this Allied force was fulfilled at 0241 local time May 7, 1945." Center, VE Day in New York. Right, a 15-year-old German soldier is taken prisoner. By the end, Germany was so desperately short of manpower that boys even younger than this one were drafted.

Hitler boasted that his creation—the German Third Reich—would live forever. Many believed him. It lasted 12 years.

MAY 7: German military leaders have surrendered to General Dwight D. Eisenhower at a school in Rheims, France. It is unconditional surrender. The war in Europe is over. It is hard to believe.

MAY 8: President Harry Truman proclaims this VE (Victory in Europe) Day. It is his 61st birthday. People are cheering and hugging and crying and partying. *Maybe they should wait—the war isn't over in the Pacific.*

MAY 11: A Japanese pilot, trained for a suicide mission, crashes into the

All over Europe, in countries that had fought for the Allies and countries that had fought for the Axis, mothers like this woman in Austria wept for joy when their sons came home or for sorrow when they did not.

The aircraft carrier *Bunker Hill* blazes up after a Japanese kamikaze (suicide) pilot has crashed into it.

WAR, PEACE, AND ALL THAT JAZZ

At the Potsdam conference, Stalin (right) talks of making Germany pay huge reparations to the Allied powers. When the Americans, headed by President Truman (center; Churchill is on the left), hear that, they threaten to leave the conference. Do you remember how Germany felt about reparations after World War I—and what came of those feelings?

aircraft carrier *Bunker Hill*. 373 Americans are killed.

June: The cities of Nagoya, Kobe, Osaka, Yokohama, and Kawasaki have been firebombed. The destruction in Japan is said to be staggering. Japan's cities are built of wood. They burn quickly. Two million buildings have been destroyed; perhaps 10 million people made homeless; hundreds of thousands are dead or injured. The Japanese warlords still won't surrender.

July: Japan's 60 largest cities have been burned. There is almost no food in Japan; people are starving. Four hundred Japanese have been arrested because they talked of surrender.

In the New Mexican desert, scientists prepare to test an atomic device. No one knows if it will work. No one knows how powerful it is. They learn on July 16: it is more powerful than anything ever before devised by humans. It was "as though the earth had opened and the skies had split," said one who was there. You, and America's other citizens, are not told of the test.

July 26: Truman, Churchill, and China's leader, Chiang Kai-shek, broadcasting from a peace conference at Potsdam, Germany, demand the unconditional surrender of Japan but assure the Japanese of a "new order of peace, security, and justice." Otherwise, they warn, there will be "prompt and utter destruction of the Japanese homeland." The Japanese premier Baron Suzuki says the proposal is "unworthy of public notice."

At Potsdam, Harry Truman tells Joseph Stalin that the United States has a new and powerful weapon. The Soviet leader shows little interest. Plans for the secret weapon, stolen by spies, sit inside the Kremlin, Russia's government center. Stalin, a dictator admired by Hitler, knows all about the bomb.

42 A Little Boy

The face of Nagasaki, Japan, 24 hours after the bomb drops. This little boy is holding his emergency ration—a ball of boiled rice.

The men of the 509th Composite Group of the 313th Wing of the 21st Bombing Command of the 20th Air Force have been carefully chosen from a group of ace pilots. All have volunteered for a special mission. No one tells them what the mission will be. But whatever it is, they know they will be flying B-29s, the big workhorse bombers that are known as superfortresses.

Right away, there is something strange about the training they get at an airfield in Utah. Instead of flying planes loaded with huge bombs, they train with a single bomb of moderate size. And they are trained to worry about storms, especially electrical storms. Then, when they are sent to the Pacific, to the island of Tinian in the Marianas group of islands, they just sit around. It is frustrating. From Tinian it is an easy flight to Japan. The other airmen on the island are flying B-29s and dropping big bomb loads on Japan's cities. The 509th is sent on training flights—over and over again. Sometimes they are allowed to drop one bomb. It isn't long before the other pilots on Tinian begin making fun of the 509th. Someone even writes a poem about them. Its last lines are:

Take it from one who knows the score,
The 509th is winning the war.

That is cruel. Everyone knows the 509th isn't winning the war; it isn't doing anything.

Meanwhile, in Washington, President Truman has come to a decision. He has called on two teams of experts: a team of scientists and a team of civilians and soldiers. They are to help him decide about the

The U.S.S. *Indianapolis* brings key components of Little Boy to Tinian. Three days later, the ship is torpedoed and sunk in the Indian Ocean.

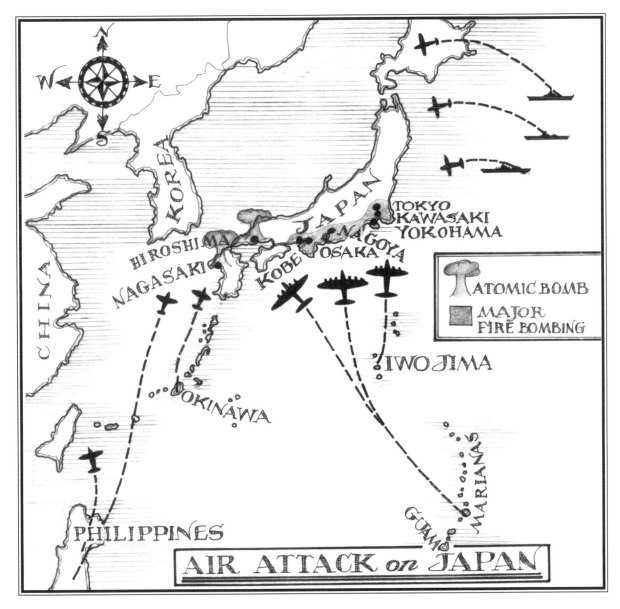

AIR ATTACK on JAPAN

new superweapon. Will it be used? Can anything be used in its place? Both teams agree: the weapon should be used. They believe it will bring the war to an end. Without it, the war could continue for 10 or more years. Military chiefs, who don't know about the secret weapon, are pressing Truman to let them invade Japan. If that happens, America can expect a million casualties; Japan might have 10 times that number.

Above, Colonel Tibbets waves from the cockpit of the *Enola Gay*, shortly before takeoff from Tinian on August 6, 1945.

Aᴮᵒᵛᴇ, Little Boy in person. The bomb was originally named Thin Man, after FDR; the Nagasaki bomb was Fat Man, after Churchill. The crew just called it "the gimmick." Bᴇʟᴏᴡ, the mushroom cloud over Nagasaki.

Colonel Paul W. Tibbets, Jr., is commander of the 509th. He has named his plane the *Enola Gay*, after his mother. In early August, a single bomb 28 inches in diameter and 10 feet long is loaded onto the *Enola Gay*. The bomb weighs four metric tons, and is nicknamed Little Boy. A similar device was exploded in the New Mexican desert, but no one knows exactly what will happen when one is dropped from an airplane. Colonel Tibbets and the others now realize that this is dangerous stuff they are about to handle. If the *Enola Gay* crashes on takeoff, as some B-29s have done, Tinian could disappear.

The plan is to drop Little Boy on the Japanese city of Hiroshima. Hiroshima has been selected because of its warmaking industries and because it is the headquarters of the 2nd Japanese Army. On August 4, more than 700,000 leaflets are dropped on Hiroshima warning that the city will be demolished. The warning is not taken seriously.

Captain William S. Parsons, of the U.S. Navy, is a surprise passenger on the *Enola Gay*. He has decided he will put the detonating parts of the bomb together after the plane is in the air. It will be safer. He doesn't know how to do that, but he has a day to learn. He learns.

At 1:45 A.M. on August 6, 1945, three B-29s take off for Japan. They will check on weather and on aircraft in the target area. The *Enola Gay* and two other B-29s follow an hour later. The night is perfect, with shining stars and a picture-book moon. As they fly, Captain Robert A. Lewis, co-pilot on the *Enola Gay*, writes a letter to his mother and father.

I think everyone will feel relieved when we have left our bomb, he writes. Later, he adds:

It is 5:52 and we are only a few miles from Iwo Jima. We are beginning to climb to a new altitude. When they are over Honshu, Japan's central island, he writes, *Captain Parsons has put the final touches on this assembly job. We are now loaded. The bomb is alive. It is a funny feeling knowing it is right in back of you.*

For most people in Hiroshima the workday begins at 8 A.M. By 8:10, factories and shops are beginning to buzz. On August 6, the entire 2nd Japanese Army is on a parade field doing calisthenics. It is a bright, sunny morning and some children can be seen outdoors playing. (Many of Hiroshima's children have been evacuated to the suburbs.) A group of middle-school students has gotten up early and already put in more than an hour's work on a fire-control project. As clocks near 8:15 A.M., the Chuo Broadcasting Station reports that three B-29s have been spotted heading for Hiroshima.

At 8:15 the bomb bay opens; Little Boy is on his way. *There will be a short intermission while we bomb our target,* Captain Lewis writes. Then he adds, in letters that scrawl wildly on the page, *My God!*

He has not been prepared for what happens. The size and fury of the explosion are greater than anything ever before created by humans. The airmen are still able to see the inferno clearly when they have put 270 miles between themselves and the target. It is a sight they will never forget.

The atomic bomb (for that is what it is) has created a fireball whose center reaches 4,000° Celsius. (Iron melts at 1,550° Celsius.) The fireball gives birth to a shock wave and then a high-speed wind. Buildings are smashed by wave and wind and burned by fire. Dust from destroyed buildings makes the city night-dark within minutes of the bombing. The wind tosses people about. Thermal rays burn their bodies. As the fireball fades, a vacuum at the blast's center pulls up dust, air, and bomb debris, creating an enormous mushroom cloud that rises into the atmosphere. Liquid rain alternates with downpours of sparks and fire. The rain is ink-black and oily. Within minutes, the 2nd Japanese Army no longer exists. Seventy-eight thousand people are dead. One hundred thousand are injured.

The atomic age has begun.

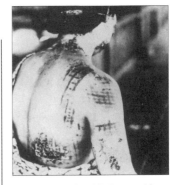

This woman's skin burned in a pattern corresponding to parts of the kimono she was wearing when the bomb went off.

Survivors in Nagasaki, the day after the explosion there. By December, the city's estimated death count was 70,000.

The destruction at Hiroshima. "It seemed impossible," said one eyewitness, "that such a scene could have been created by human means."

43 Peace

VE Day had been a rehearsal. On VJ Day, Americans everywhere went wild.

The Americans demand surrender—otherwise, they tell the Japanese, "they may expect a rain of ruin from the air." The Japanese do not respond.

On August 8 Russia enters the war against Japan. Russian forces attack Japanese armies in Manchuria and Korea. For some Japanese leaders this is more threatening than the bomb.

On August 9 a second atomic bomb is dropped. This one hits the port of Nagasaki. The nightmare of fire, wind, rain, and radiation is repeated. Has this war driven sane people to act insanely?

In Japan, the warlords and government ministers meet in stormy sessions to decide the fate of their people. They are split down the middle; most of the warriors vote to continue the fight, although everyone knows that may mean the destruction of the Japanese people. As Emperor Hirohito described it:

> There was no prospect of agreement no matter how many discussions they had....I was given the opportunity to express my own free will for the first time.

The emperor decides for his people.

From Tokyo, cables are sent to Washington, London, Moscow, and Chungking. Japan will accept the terms of the Potsdam Declaration, with one request: that the emperor remain as head of state.

"The time has come," said Emperor Hirohito (center, with General Tojo, right), "when we must bear the unbearable."

As soon as Harry Truman hears the news he calls Eleanor Roosevelt. Mrs. Roosevelt is now America's ambassador to the United Nations. President Truman tells her that he wishes her husband were alive to accept the peace proposal.

In Japan, some army officers break into the imperial palace, set fire to the home of the prime minister, and attempt to stop a broadcast of the emperor's words. They fail, and, along with the war minister, commit suicide in the public square.

The emperor, who wants to be sure that his words are exactly right, has recorded them in advance. The actual broadcast is taking place elsewhere.

It is August 15, 1945, and Hirohito's voice is heard in the first public speech an emperor has ever made. He asks the Japanese people to accept the coming of peace.

Above, General MacArthur (behind table, back to camera) watches as General Yoshijiro Umezu signs Japan's surrender.

New York City on VJ Day. Times Square is closed to traffic; 2 million people squeezed into the area are hugging and cheering.

Chronology of Events

1905: President Theodore Roosevelt's niece Eleanor marries her fifth cousin Franklin D. Roosevelt

1907: Oklahoma becomes the 46th state

1912: New Mexico and Arizona become the 47th and 48th states

1914: war breaks out in Europe between the Entente countries (Britain, France, Italy, Russia) and the Central Powers (Germany, Austria-Hungary, Turkey)

1916: Woodrow Wilson is reelected president

Apr. 1917: the U.S. enters the war on the side of the Entente powers

Nov. 1917: in Russia, the revolutionary communist Bolshevik Party overthrows a more moderate regime and takes Russia out of the war

1918: a worldwide influenza epidemic kills nearly 500,000 Americans

1918: Armistice Day, November 11: the war is over; Germany accepts Wilson's Fourteen Points as the basis for peace negotiations

1919: Babe Ruth joins the New York Yankees and helps restore baseball's popularity after the Chicago Black Sox World Series scandal the same year

1919: at the Paris peace conference, Germany is humiliated and must pay punitive reparations; the Allies write a charter for a League of Nations

1919: Wilson asks Americans to support the League, but a stroke makes him hardly able to function

1919: the 18th Amendment to the Constitution prohibits making or selling alcohol in the U.S.

1920: the 19th Amendment gives women the vote

1920: Republican Warren G. Harding, advocating a "return to normalcy," becomes 29th president

1920: in a "red scare," thousands of immigrants are arrested on suspicion of being communist

1920: KDKA, the nation's first commercial radio station, begins broadcasting out of Pittsburgh

1920: the Census shows that for the first time more than half the United States population is urban

1921: Franklin D. Roosevelt is crippled by polio

1922: Harding ignores government corruption, which culminates in the Teapot Dome scandal

1922: Louis Armstrong leaves New Orleans to play for King Oliver's Creole Jazz Band in Chicago

1922: the era of the Harlem Renaissance in the arts among black Americans begins with the publication of Claude McKay's *Harlem Shadows*

1923: When Harding dies suddenly, Vice President Calvin Coolidge becomes 30th president; he is reelected the following year

1925: F. Scott Fitzgerald publishes *The Great Gatsby*

1925: a high-school teacher, John Scopes, is tried and convicted in Tennessee for teaching evolution

1926: scientist Robert Goddard launches the first successful rocket powered by liquid fuel

1927: Charles Lindbergh makes the first solo flight across the Atlantic in the *Spirit of St. Louis*

1927: the first talking motion picture, *The Jazz Singer*

1927: in *Nixon* v. *Herndon*, the Supreme Court rules that a Texas law preventing blacks from voting in primary elections is unconstitutional

1927: Babe Ruth hits 60 home runs

1928: Walt Disney's Mickey Mouse makes his first appearance in *Steamboat Willie*

1928: Herbert Hoover becomes 31st president, defeating Democrat Al Smith in a landslide

1928: FDR is elected governor of New York

1929: the stock market crashes in October and hastens a worldwide economic depression

1931: Japan invades Manchuria

1932: the Bonus Army of war veterans marches to Washington, asking for early payment of war benefits

1932: Mildred "Babe" Didrikson wins three track-and-field medals at the Olympic Games

1932: Depression: one in four workers in America —12 million people—are unemployed

1932: Democrat Franklin D. Roosevelt pledges to beat the Depression and becomes 32nd president

1933: Adolf Hitler, leader of the Nazi Party, becomes chancellor of Germany

1933: Frances Perkins becomes secretary of labor and the first woman cabinet member

1933: the first "100 Days" of FDR's presidency: Congress creates more agencies and passes more laws to relieve poverty than ever before

1933: the 21st Amendment repeals Prohibition

1934–37: after three years of drought, the farms of the Plains have become the Dust Bowl; 3 million victims migrate west in search of work and land

1935: Social Security Act introduces taxes to pay for unemployment insurance and old-age pensions

1936: Jesse Owens wins four gold medals in track events at the Olympic Games in Berlin

1938: Germany annexes Austria

1939: all German Jews made to wear yellow stars of David sewn onto their clothes for identification

1939: Germany occupies Czechoslovakia

1939: Stalin signs a nonaggression pact with Hitler

1939: Germany invades Poland; Allies declare war

1939: physicist Albert Einstein helps convince FDR to fund the Manhattan Project to build atom bomb

1940: Germany rapidly overwhelms Norway, Finland, Denmark, Holland, Belgium, and France

1940: FDR is first president elected to three terms

1940: Germany tries to bomb Britain into submission

Jun. 1941: Hitler ignores the Nazi–Soviet Pact, invading the Soviet Union; Russia joins the Allies

Sep. 1941: the siege of Leningrad begins

1941: FDR creates the Lend-Lease program to supply the Allies with weapons and raw materials

1941: Germany pushes deep into Russia and Ukraine

Dec. 1941: Japan bombs the U.S. fleet in Pearl Harbor, Hawaii; America enters the war

Dec. 1941–Jan. 1942: Japan invades Thailand, Malaya, Borneo, Burma, the Philippines, Hong Kong, Indonesia; captures Midway and Wake islands

1942: FDR orders 112,000 Japanese Americans into internment camps for the war's duration

May 1942: heavy losses for both U.S. and Japan in battle of Coral Sea; first setback for Japan

May 1942: U.S. surrenders in Philippines

June 1942: U.S. defeats Japan at battle of Midway

Aug. 1942: U.S. defeats Japan at Guadalcanal

1942: U.S. forces join British and French to fight Germany and Italy in North Africa

1942: the Nazis begin systematic murder of Jews and others in concentration camps

1942: the Allies capture Enigma, a code machine, and begin to crack German secret codes

Feb. 1943: the Soviet Red Army defeats Germany at Stalingrad and begins to push Germany back

June 1943: Allies win battle of Atlantic

July 1943: Allied forces invade Sicily

Sep. 1943: Italian army surrenders

Nov. 1943–Feb. 1944: U.S. takes back numerous Pacific islands from Japanese

June 1944: Allied forces enter Rome

June 1944: combined U.S., British, and Canadian forces invade Normandy, France, on D-Day, June 6

Nov. 1944: FDR wins fourth term!

Jan. 1945: the U.S. returns to the Philippines

Jan. 1945: the Allies defeat Germany in Battle of the Bulge in the Ardennes mountains in Belgium

Feb. 1945: Roosevelt, Churchill, and Stalin meet at Yalta in the Crimea

Feb. 1945: Allies firebomb Dresden, Germany

Feb. 1945: U.S. troops land on island of Iwo Jima

Apr. 1945: Hitler commits suicide in Berlin

Apr. 1945: President Roosevelt dies; Harry S. Truman sworn in as 33rd president

Apr. 1945: Soviet troops enter Berlin

Apr. 1945: U.S. troops liberate 32,000 prisoners in Dachau concentration camp, Germany

May 1945: Germany surrenders; VE Day, May 8

June 1945: charter established for United Nations

July 1945: Truman, Churchill, and Stalin meet at Potsdam in Germany to plan for peace in Europe

Aug. 1945: atomic bombs dropped on Japanese cities of Hiroshima and Nagasaki

Aug. 1945: Japan surrenders; VJ Day, August 15

More Books to Read

Here are some books that tell about life during the period discussed in this book. Some were written for grownups—but they are not too hard to read, they tell wonderful stories, and they are the kind of books you will always remember. Besides these books, there is something else you should search out if you don't know about it: Cobblestone, which is a history magazine for young people. Each issue focuses on a subject in history and back issues can be found in most libraries or ordered from Cobblestone Publishing, 30 Grove Street, Peterborough, N.H. 03458.

F. Scott Fitzgerald, *The Great Gatsby,* Scribner's, 1925. The best American novel of the 1920s. This story of crazy pleasure seekers and unrequited love is not very long, and it will make you laugh and cry at the same time. Then try some of Fitzgerald's short stories, which are quite easy to read.

Anne Frank, *The Diary of a Young Girl,* Modern Library, 1958. In Dutch, the language Anne Frank wrote in, her diary is called *The House Behind,* because she and her family, who were Jews, hid from the Nazis in a few secret rooms inside a Dutch family's Amsterdam home. This book may make you cry, but you will love Anne and be amazed at what a marvelous writer she was at the age of 14.

Ernest Hemingway, *A Farewell to Arms,* Scribner's, 1929. An enthralling novel that draws on the author's own experience as a soldier in the Great War.

John Hersey, *Hiroshima,* Knopf, 1946. This little book—it was originally an article in *The New Yorker* magazine—is a clear, sympathetic telling of the dropping of the first atomic bomb and its devastating effect on Hiroshima and its people. You should also try another very good book by John Hersey about the war called *A Bell for Adano* (Knopf, 1944).

Gloria Houston, *Littlejim,* Philomel, 1990. Littlejim Houston lives in the Appalachian mountains in North Carolina during the Great War. He's the best student at his school and works hard on the farm—but Littlejim's father wants his son to be a logger and hunter like him, and he thinks being good at studying is sissy. Then Littlejim enters a big essay contest on "What it means to be an American"; what will his father think if he wins?

Jeanne Wakatsuki Houston and James D. Houston, *Farewell to Manzanar,* Houghton Mifflin, 1973. In 1942, when the author was seven years old, she and all her family, along with thousands of other Japanese Americans, were sent suddenly to Manzanar internment camp in the desert of California's Owens Valley. Her father was sent even farther away; it was nine months before he was allowed to rejoin them. This book describes vividly the Wakatsukis' humiliations and hardships.

Zora Neale Hurston, *Their Eyes Were Watching God,* 1937 (several editions available). A terrific and very influential story about growing up poor and black in the South early in this century.

Jerome Lawrence and Robert E. Lee, *Inherit the Wind,* Bantam, 1969. This is a thrilling play (also made into a good movie that you might be able to find on video) which is a fictionalized dramatization of the Scopes "monkey trial" of 1925.

Lois Lenski, *Strawberry Girl,* Lippincott, 1945. Birdie Boyer is ten when her family moves to the Florida backwoods to grow sugarcane, oranges, and strawberries. She finds out that it's hard dealing with drought and frost, but it's much harder to cope with unfriendly neighbors: the Slaters. This excellent story is all about how the Boyers farm, play, go to school—and deal with the Slaters.

Sonia Levitin, *Journey to America; Silver Days,* Atheneum, 1970, 1989. In 1938 Lisa Platt's father left Berlin for America in the middle of the night. Lisa, her sisters, and her mother have to wait until he has made enough money to send for them. But life for Jews in Hitler's Germany becomes more and more scary; finally the Platts escape to Switzerland—where they wait and wait, with no money and not enough to eat, for a letter to arrive from America. It wasn't easy even when they got here. In *Silver Days,* the sequel, the author tells about the difficulties of being poor Jewish refugees in wartime California.

Cornelius Ryan, *The Longest Day,* Simon &

Schuster, 1959. A really exciting retelling of the story of D-Day and the invasion of Normandy.

John Steinbeck, *The Grapes of Wrath,* Viking, 1939. The Joad family are sharecropper victims of the Dust Bowl in the 1930s. Piling everything they own onto their decrepit car, they set off to look for work picking fruit and vegetables in California. This famous book helped bring home the plight of the "Okies" to the rest of the country. It is a great read.

Mildred D. Taylor, *Let the Circle Be Unbroken; Mississippi Bridge; The Road to Memphis; Roll of Thunder, Hear My Cry,* all published by Dial. Cassie Logan and her family live in rural Mississippi in Jim Crow times when black people who got "uppity" could be risking their lives. Mildred Taylor's four books are about the racism of the Deep South in the 1920s and '30s and how it affects the Logans at different times during Cassie's childhood and youth. They are very well written and sometimes hard to take.

Jade Snow Wong, *Fifth Chinese Daughter,* U. of Washington Press (first pub. 1945). Jade Snow Wong's parents owned a small overalls factory in San Francisco in the 1920s, when she was born. Until she was about ten, Jade Snow led a very sheltered, Chinese life—looking after little sister, learning to cook rice perfectly, and discovering that grownups thought boys were more important than girls. Jade Snow decides to work hard in school; eventually she must convince her parents to allow her to be the first girl in her family to go to college. This is a splendid true story of the growing up of not just one Chinese daughter, but a whole family.

Index

Picture Credits

AR: U.S. Army
FDR: Franklin Delano Roosevelt Library, Hyde Park, NY
LOC: Library of Congress

MCNY: Museum of the City of New York
NA: National Archives
NBL: National Baseball Library
NPG: National Portrait Gallery,

Smithsonian Institution
NV: U.S. Navy
OWI: Office of War Information
OSA: Office of the Secretary of Agriculture

USIA: U.S. Information Agency
WPA: Works Progress Administration
WRA: War Relocation Authority

Cover: Scala/Art Resource, New York; **5**: Joy Hakim; **6 (top)**: FDR; **6–7**: NA/OSA; **7**: Freelance Photographers Guild; **8**: LOC; **9 (top)**: detail, Nathaniel Currier, *The Way They Go to California*, 1849, Oakland Museum; **9 (left)**: NA; **9 (top, middle right)**: Bowdoin College Museum of Art, Brunswick, Maine; **9 (bottom right)**: Museum of Fine Arts, Boston/NPG; **10 (top)**: U.S. Capitol Historical Society and National Geographic Society; **10 (bottom)**: LOC; **11**: NA/WPA; **12**: Jane Bradick, *View of West Front of Monticello and Garden*, 1825, private collection; **13–14**: NA; **15 (left)**: Wide World; **15 (right)**: *New York Times*; **16**: Culver; **17**: NA; **19 (top)**: Imperial War Museum; **19 (bottom left)**: Harding, *Brooklyn Eagle*, 1919; **19 (bottom right)**: *Chicago News*, 1919; **20 (top)**: Harris & Ewing; **20 (bottom left)**: Holmes Papers, Harvard Law School Library; **20 (bottom right)**: LOC; **21 (top)**: National Library of Medicine, Bethesda, MD; **21 (bottom), 22 (top)**: Sy Seidman; **22 (bottom)**: Bettmann; **23**: LOC; **24**: Marcel Duchamp, *Nude Descending a Staircase, No. 2*, 1912, Philadelphia Museum of Art, Louise and Walter Arensberg Collection; **25 (bottom), 26 (top)**: MCNY; **26 (bottom)**: UPI/Bettmann; **27, 28**: MCNY; **29 (top)**: Brown Brothers; **29 (bottom), 30**: LOC; **31**: National Museum of American History, Smithsonian Institution; **32 (bottom left)**: Brown Brothers; **32 (bottom right)**: LOC, French Collection; **33 (top)**: Douglas County Museum photograph, Roseburg, Oregon; **33 (bottom left)**: Bettmann; **33 (bottom right)**: League of Women Voters of the City of New York; **34 (top)**: Culver; **35 (top left)**: Brown Brothers; **35 (top right)**: Culver; **35 (bottom left)**: Wide World; **35 (bottom right)**: Museum of Modern Art/Scala; **36 (top left, bottom right)**: UPI; **36 (right)**: Morgan, *Philadelphia Inquirer*, March 13, 1919, New York Public Library; **37, 38**: LOC; **39 (top)**: Underwood & Underwood; **39 (bottom)**: Brown Brothers; **40 (top)**: Ellison Hoover, *Life*, March 6, 1924; **40 (middle, bottom)**: Phillips Collection, Washington, D.C.; **41**: Culver; **42 (top left)**: Brown Brothers; **42 (top right)**: UPI; **42 (bottom)**: LOC; **43, 44 (left)**: Culver; **44 (right)**: Brown Brothers; **45**: UPI; **46 (top)**: Alfred Stieglitz, 1918, Metropolitan Museum of Art; **46 (bottom left)**: Bettmann; **46 (bottom right)**: Brown Brothers; **47 (top left)**: LOC; **47 (top right)**: Paul MacFarlane–Mary Broderick; **47 (bottom)**: Harris Lewine; **48**: Culver; **49 (left)**: Bronx County Historical Collection, Bronx, NY; **49 (right)**: Brown Brothers; **50 (inset)**: Culver; **50 (bottom)**: Bettmann; **50 (bottom right)**: UPI; **51 (top left)**: NBL; **51 (top middle)**: collection of Phil Dixon; **51 (top right)**: James Bell; **51 (bottom)**: Mike Anderson; **52 (top)**: NBL; **52 (middle)**: National Baseball Hall of Fame; **52 (bottom)**: *Kansas City Call*; **53 (top)**: Don Rogosin; **53 (middle)**: National Baseball Hall of Fame; **53 (bottom)**: Mark Rucker; **54 (top)**: Bettmann; **54 (bottom)**: *New York Daily News*; **55 (bottom)**: Malcolm F. J. Burns; **56**: collection of Leonard V. Huber; **57**: Archibald Motley, Jr., *Blues*, 1929, collection of Archie Motley and Valerie Gerrard Browne; **58 (top, middle)**: LOC; **58 (bottom)**: Historical Pictures Service, Chicago; **59**: Ernest R. Smith; **60 (top)**: Ralph Crane, *Life*, *The Razor's Edge* (20th Century Fox, 1946) ; **60 (bottom)**: *The Little Colonel* (1935); **61 (left)**: Wide World; **61 (right)–64**: Clark University, Worcester, MA; **65**: Minnesota Historical Society; **66**: NA; **66 (top inset)**: Culver; **66 (bottom inset)**: Brown Brothers; **69, 70 (top)**: Culver; **70 (middle)**: LOC; **70 (bottom)**: Granger; **71 (top)**: *New York Times*, November 23, 1922, American Social History Project; **71 (middle)**: Bettmann; **71 (bottom)**: NA/Records of the Forest Service; **72**: Granger; **73, 74 (top)**: Bettmann; **74 (bottom), 75**: LOC; **76**: Granger; **77**: Milton "Pete" Brooks, *Detroit News*, July 1930; **78 (top)**: Ivan E. Prall; **78 (bottom), 79**: LOC; **80**: Mrs. Henry Rhoades; **80 (inset)**: NA/Records of Soil Conservation Service; **81 (top)**: LOC; **81 (bottom)**: J. Lee, March 30, 1933, Neg. 20102, Special Collections Division, University of Washington Library; **82**: Granger; **83 (top)**: Bettmann; **83 (bottom)**: NA; **84 (top)**: Margaret Bourke-White, *Life*; **84 (bottom)**: Wide World; **85 (top)**: *Literary Digest*, November 21, 1931, New York Public Library; **85 (bottom)**: FDR; **87 (top)**: LOC; **87 (bottom left, right)–89**: FDR; **90**: LOC; **91**: FDR; **92 (top left)**: Bettmann; **92 (bottom left, right)**: FDR; **93 (left)**: Bettmann; **93 (top right)**: FDR; **93 (bottom right)**: AP/Wide World; **94 (left)**: Thomas McAvoy, *Life*, © 1939 by Time, Inc.; **94 (top right)**: Historical Pictures Service, Chicago; **94 (bottom right)**: Bettmann; **95 (left)**: LOC; **95 (right)**: Bettmann; **96, 97**: FDR; **97 (bottom)**: LOC; **98, 99**: Bettmann; **99 (bottom left)**: John Tinney McCutcheon, *Chicago Tribune*, 99 (bottom right): LOC; **100 (top)**: UPI; **100 (bottom)**: LOC; **101 (top)**: Bettmann/UPI Newsphoto; **101 (bottom)**: Miguel Covarrubias, *The Inauguration of FDR*, 1933, *Vanity Fair*, NPG; **102 (top left , right)**: LOC; **102 (bottom right)**: NEA, Inc./Culver; **102 (bottom left)**: FDR; **103**: UPI/Bettmann; **104 (top)**: Peter Arno, *The New Yorker*, 1936; **104**

(bottom): LOC; **105 (top)**: Granger; **105 (bottom)**: Wide World; **106 (left)**: Camera Press Pix; **106 (right)**: LOC; **107 (top)**: European Picture Service; **107 (bottom right)**: collection of Kenneth W. Rendell; **108 (bottom left)**: UPI; **108 (bottom right)**: Wide World; **109 (top left)**: John Erickson, Edinburgh; **109 (top right)**: Otto Hagel; **109 (bottom right)**: FDR; **109 (bottom middle)**: Wide World; **110 (top)**: L. M. Muller; **110 (bottom)**: Hulton Picture Service; **112 (top)**: NA; **113 (top left)**: Archives of Yivo, Institute for Jewish Research; **113 (top right)**: Peter Hunter, Rijksinstituut voor Oorlogsdocumentatie, Amsterdam; **113 (bottom)**: © Leni Sonnenfeld, University of the State of New York; **114**: Wiener Library, London; **115 (top left, right)**: NA/AR; **115 (middle)**: Henry Beville; **115 (bottom)**: NA/AR; **116 (top)**: Granger; **116 (bottom)**: Bettmann/UPI; **117**: courtesy Dr. Liane Reif-Lehrer; **118 (top)**: General Electric; **118 (bottom)**: LOC; **119 (top left)**: Los Alamos Historical Museum Photo Archives; **119 (top middle, top right, middle)**: Los Alamos National Laboratory and National Atomic Museum; **119 (bottom)**: Argonne National Laboratory and National Atomic Museum; **120 (top)**: NA/USIA; **120 (bottom)**: courtesy *Washington Star*; **121 (top, middle)**: AP/Wide World; **121 (bottom)**: U.S. Air Force; **122 (top)**: Black Star; **123 (top left)**: NA; **123 (top right)**: CTK, Centralna Agencja Fotograficzna; **123 (bottom right)**: Bettmann/UPI; **124 (top left)**: NA/OWI; **124 (top right)**: International War Museum; **124 (bottom)**: Netherlands State Institute for War Documentation; **125 (top)**: NA/Foreign Records Seized; **125 (bottom)**: collection of U.S.S. *Arizona* Memorial, National Park Service; **125 (left inset)**: NV/Captured Japanese Records; **125 (right inset)**: Bettmann; **126 (bottom)**: UPI; **127 (bottom)**: George Stock, *Life*; **128 (top)**: Fox Photo; **128 (bottom)**: *Detroit News*; **129 (top)**: LOC; **129 (bottom)**: Roger-Viollet, Paris; **130 (top)**: Bildarchiv Preussischer Kulturbesitz, Berlin (West); **130 (bottom)**: NA/USIA/New York Times Paris Bureau Collection; **131 (bottom right)**: Gillon Photo Agency; **132 (top left)**: Bundesarchiv, Koblenz; **132 (middle left)**: Vera Inber; **132 (bottom left, right)**: Sovfoto; **133**: NA/NV; **134 (left)**: Bison Picture Library, London; **134 (right)**: NA/AR; **135 (top right)**: NA/Coast Guard; **135 (top right, bottom)**: NA/NV; **136 (top)**: NA/USIA/New York Times Paris Bureau Collection; **136 (bottom)**: collection of Kenneth W. Rendell; **137 (top)**: NA/NV; **137 (bottom left)**: UPI; **137 (bottom right)**: NA; **138 (top)**: NA; **138 (bottom)**: American Red Cross; **139 (top right)**: German Bundesarchiv; **139 (middle right)**: NA/Coast Guard; **139 (bottom right)**: NA/AR; **139 (top left)**: Bettmann/UPI; **139 (bottom left)**: NA/Coast Guard; **140 (bottom)**: NA/NV; **142, 143 (top)**: NA/WRA; **143 (bottom)**: NA; **144 (top, middle, bottom)**: from Miné Okubo, *Citizen 13660*, Columbia University Press, 1946; **145**: NA/AR; **146 (top)**: NA/WRA; **146 (bottom)**: photograph by Toyo Miyatake, San Gabriel, California; **147 (top)**: NA/NV; **147 (bottom)**: NA/AR; **148 (bottom), 149 (top)**: NA/Marine Corps; **149 (bottom)**: NV/Combat Art Division; **150 (left)**: Joy Hakim; **151 (top)**: NA/NV; **151 (bottom)**: NA/Marine Corps; **152 (top)**: NA/War Production Board; **152 (left)**: NA; **152 (bottom middle, top right, bottom right)**: © Planeta Publishers, Moscow; **153 (left)**: NA/OWI; **153 (right)**: NA; **154 (top left)**: NA/AR; **154 (top right)**: NA/AR; **154 (bottom left)**: Ullstein-Birnback; **155 (top middle)**: Bettmann; **155 (bottom)**: BPK; **156**: NA; **157 (left)**: NA/AR; **157 (right), 158 (top)**: NA/Coast Guard; **159 (bottom right)**: Imperial War Museum; **159 (right)**: Robert Capa Archive, International Center of Photography, NYC; **161 (top)**: Wide World; **161 (middle)**: UPI; **161 (bottom)**: NA; **162**: NA/AR; **163 (top)**: FDR; **163 (top right)**: *Arizona Republic*, 1943; **163 (middle)**: NA/AR; **163 (bottom)**: Dmitri Baltermants, Moscow; **164 (top)**: NA; **164 (middle)**: Joe Rosenthal, AP, NA/NV; **164 (bottom)**: NA; **165 (bottom)**: Ullstein Bilderdienst, Berlin (West); **166, 167**: FDR; **168 (top)**: LOC; **168 (bottom)**: Harry S. Truman Library; **169 (top left)**: Bettmann; **169 (top right)**: Stan Cohen; **169 (bottom)**: Harry S. Truman Library; **170 (top)**: Bettmann; **170 (bottom), 171 (top)**: FDR; **172 (top)**: Wayne Miller; **172 (bottom)**: NA; **173**: Bettmann; **174**: Wide World; **175 (top left)**: Wide World; **175 (top right)**: Keystone, Hamburg; **175 (middle left)**: Hulton Deutsch Collection, London; **175 (middle)**: NA/AR; **175 (bottom right)**: Ivan Shaquin, TASS from Sovfoto; **176 (top middle)**: Bettmann; **176 (top right)**: Johnny Florea, *Life*; **176 (middle left)**: Frederico Patellani, Milan; **176 (bottom right)**: NA; **177**: LOC; **178**: Yosuke Yamabata; **180 (top left)**: Bettmann; **180 (top right)**: NA/OWI; **180 (bottom)**: NA; **181 (top)**: NA/Corps of Engineers; **181 (middle)**: Eiichi Matsumoto; **181 (bottom)**: UPI/Bettmann; **182 (top)**: Eastern Washington State Historical Society, Spokane; **182 (bottom)**: NA; **183 (top)**: Carl Mydans, *Life*, **183 (bottom)**: FDR; **183 (inset)**: Alfred Eisenstaedt, PIX, Inc.; **191**: Jack Delano, LOC/Farm Security Administration.

Tobacco farmers keep a sense of humor despite a depression.

A Note From the Author

It was the best of times, it was the worst of times, it was the age of wisdom, it was the age of foolishness....

Charles Dickens was thinking about the 18th century when he began a famous novel with those words. But he might have been describing the first half of the 20th century.

The best of times? The worst of times?

Which was it? It was both.

Take that bomb, for instance. It was horrible to think of its hellish, destructive force. But there was another side to atomic energy. Wisely used, it held great hope for the future.

Wisdom and foolishness in modern times? You bet. And sometimes it was hard to tell them apart. Should the bomb have been dropped? The answer was obvious: of course not.

It was outrageous, and barbaric, to bomb children and other civilians with any kind of bomb. The bombs dropped on Hiroshima and Nagasaki were the terrible outcome of a misguided idea— strategic bombing (which meant bombing of any targets, including civilian, that might aid the war). Strategic bombing was first tried in Spain in 1939 by Fascist forces. Hitler took the idea further when he bombed Britain. The Allies went still further.

But there was another question that had to be asked. What would you have done if you were Harry Truman? If you had a weapon that would end the war quickly (and save lives on both sides), would you have used it? Of course you would. In 1945, hardly anyone even thought of hesitating.

There was a lesson in all this. It isn't easy to be a citizen in a democracy. It takes work: reading, listening, and questioning. And yet, after the war, some people were still crying out for simple solutions. They were looking for leaders with slogans, not thinking explanations. They weren't willing to take time to become responsible citizens. A good citizen needs to be well informed, to get involved, and to know history.

Why history? Because it helps us make judgments. And history shows us that ideas set in motion sometimes head in unforeseen directions. They may lead to places we wish they wouldn't go.

The people in Germany who followed Hitler didn't expect his ideas to take them into the sewers of evil and destruction. But they didn't examine those ideas carefully. They didn't protest while they still had a chance. They didn't consider other people's feelings. They listened to simplistic messages, and they let others think for them.

The people in the United States who were racists and bigots didn't usually mean to hurt others. But they didn't do much thinking. History tells us that disgusting, ugly ideas are sometimes accepted by ordinary (but unthinking) people.

Did we learn anything from the history of the first half of the 20th century?

Yes. We learned of the horror and stupidity of totalitarian rule. (In totalitarian governments, a powerful leader is usually in league with the military, the police, and sometimes with business interests. Once they seize power, the ordinary citizen doesn't stand a chance.)

We'd learned of the waste of war. What if all those 20th-century children who died of violence had lived? What might they have accomplished?

And what of the waste of resources? Destroying a million dollars' worth of city buildings and streets usually costs the other side a million dollars in planes, bombs, and equipment. (Imagine if that money were spent to make life better for people.)

We'd learned something else: in a time of crisis, the best in people often appears. There was much real heroism during those years of war and depression. People worked together, and sacrificed, and felt good about that, and about each other. Is there some way to bring that sense of shared emergency to today's problems? How can we work together to make our cities safe? To improve our schools? To help all our citizens pursue happiness?